an inconvenient truth

an inconvenient truth

the crisis of global warming

AL GORE

Originally published in another form by Rodale Inc., 2006
This revised edition first published in 2007 by Viking, a division of Penguin Young Readers Group and Rodale Inc.
Rodale Inc., 733 3rd Avenue, New York, New York 10017

Viking
Published by Penguin Group
Penguin Young Readers Group, 345 Hudson Street, New York, New York 10014, U.S.A.
Penguin Group (Canada), 90 Eglinton Avenue East, Suite 700, Toronto, Ontario, Canada M4P 2Y3
(a division of Pearson Penguin Canada Inc.)
Penguin Books Ltd, 80 Strand, London WC2R 0RL, England
Penguin Ireland, 25 St Stephen's Green, Dublin 2, Ireland (a division of Penguin Books Ltd)
Penguin Group (Australia), 250 Camberwell Road, Camberwell, Victoria 3124, Australia
(a division of Pearson Australia Group Pty Ltd)
Penguin Books India Pvt Ltd, 11 Community Centre, Panchsheel Park, New Delhi – 110 017, India
Penguin Group (NZ), 67 Apollo Drive, Mairangi Bay, Auckland 1311, New Zealand
(a division of Pearson New Zealand Ltd)
Penguin Books (South Africa) (Pty) Ltd, 24 Sturdee Avenue, Rosebank, Johannesburg 2196, South Africa

Penguin Books Ltd, Registered Offices: 80 Strand, London WC2R 0RL, England

Adapted for young readers by Jane O'Connor

10 9 8 7 6 5 4 3 2 1

LIBRARY OF CONGRESS CATALOGING-IN-PUBLICATION DATA
Gore, Albert, date–
An inconvenient truth / by Al Gore.
p. cm.
Adaptation of: An inconvenient truth / by Al Gore. 2006.
Includes bibliographical references and index.
ISBN 978-0-670-06271-3 (hardcover) — ISBN 978-0-670-06272-0 (pbk.) 1. Global warming—Juvenile literature. 2. Greenhouse effect, Atmospheric—Juvenile literature. 3. Climatic changes—Juvenile literature. 4. Environmental protection—United States—Juvenile literature. I. Title.
QC981.8.G56G668 2006
363.73874—dc22
2006103242

Printed in the U.S.A.
Set in Old Style 7 and Akkurat
Book design by Jim Hoover and mgmt. design

Caney Fork River,
Carthage, Tennessee, 2006.
Photograph by Tipper Gore

Al and Tipper Gore one month before the birth of their first child, Karenna, on the Caney Fork River, Carthage, Tennessee, 1973

For my beloved wife and partner, Tipper,
who has been with me for the entire journey

CONTENTS

Introduction

MY CHILDHOOD WAS spent in two places. I grew up half in the city and half in the country. My father was a senator from Tennessee and worked in Washington, D.C., so that's where I went to school. But summers were spent on our family's farm in Carthage, Tennessee. I'd go from living in a small eighth-floor apartment whose windows looked out on concrete parking lots and buildings to a sprawling farm with animals, sunlight, open sky, and the sparkling water of the Caney Fork River.

Over time, I came to love my days on the farm more and more: the soft grass, rustling trees, cool lakes. I often walked every inch of the farm with my father. He taught me many lessons about nature on our walks. He showed me how to help preserve the soil by using rocks or branches to stop rivulets of rainwater from washing away the topsoil. These days, when I walk on what is now my farm with my children and grandchildren, I teach the same lessons my father taught me about our duty to care for the land.

I first learned about the Earth's vulnerability to human hands from my mother. When I was fourteen, she read a book called *Silent Spring* by Rachel Carson. She thought its message that human civilization now had the power to seriously harm the environment was so important that she read it to my sister and me. The book's lessons made a huge impression on us. The way we thought about nature and the Earth was never the same again.

Then in 1968 when I was in college, I learned even more about how the natural world I loved so deeply was in peril. A great teacher of mine at Harvard, Dr. Roger Revelle, opened my eyes to the problem of global warming. Like all great teachers, he influenced the rest of my life. He shared with me and my classmates what was happening to the atmosphere of the entire planet, and how that enormous change was being caused by human beings. During my twenty-four years in government and now as a private citizen, I have always worked to alert people to the dangers of global warming and help figure out how to stop it.

Global warming may not seem like one of our biggest dangers, but it is. Science has now proven beyond a doubt that the Earth's climate is changing, and changing much faster than originally feared. Because of it, we are witnessing such awful results as Hurricane Katrina in 2005—results that leave so many people homeless and cities devastated.

Global warming is not caused by natural forces beyond our control. No asteroid has struck; the Earth is not moving closer to the sun. Human beings are the main cause of the problem. And so it is our responsibility to fix it.

Your generation has grown up much more aware of environmental problems than mine. You already understand that our relationship to nature is not a relationship of "us" and "it." You know that we are all part of the same ecosystem, that we are all in it together.

Yes, this is a crisis. However it is not hopeless. I like the fact that in Chinese, the word "crisis" is made up of two characters: 危机. The first one means "danger." But the second means "opportunity."

I hope the readers of this book will see the opportunity to live differently and to make changes that will help end the climate crisis.

Al, his sister Nancy, and their parents on the Caney Fork River, Carthage, Tennessee, 1951

chapter one

Our Changing Planet

THIS IS THE first picture most of us ever saw of the Earth from space. It was taken on Christmas Eve, 1968, by one of the astronauts aboard the Apollo 8 spacecraft. For the very first time, a manned spacecraft was circling the moon (whose surface you see at the bottom of the picture), scouting for possible landing sites for future missions. The work of Apollo 8 led to the amazing moment the following summer when Apollo 11 touched down on the moon.

The entire time Apollo 8 traveled around the far side of the moon, the Earth disappeared from view. The astronauts were alone in the black void of space, with no radio contact, as NASA had expected. Total isolation.

Then, as radio contact was reestablished, the crew looked up and saw this spectacular sight.

Author Archibald MacLeish wrote a day later on Christmas, "To see the Earth as it truly is, small and blue and beautiful in that eternal silence where it floats, is to see ourselves as riders on the Earth together. . . ."

This famous photo became known as "Earth Rise." It literally changed our view of the planet, and in doing so, it did something much more important. It changed our attitude toward the Earth as well. We saw its beauty in a new way. During the next few years in our country, the Clean Air Act, the Clean Water Act, and the first Earth Day all came about.

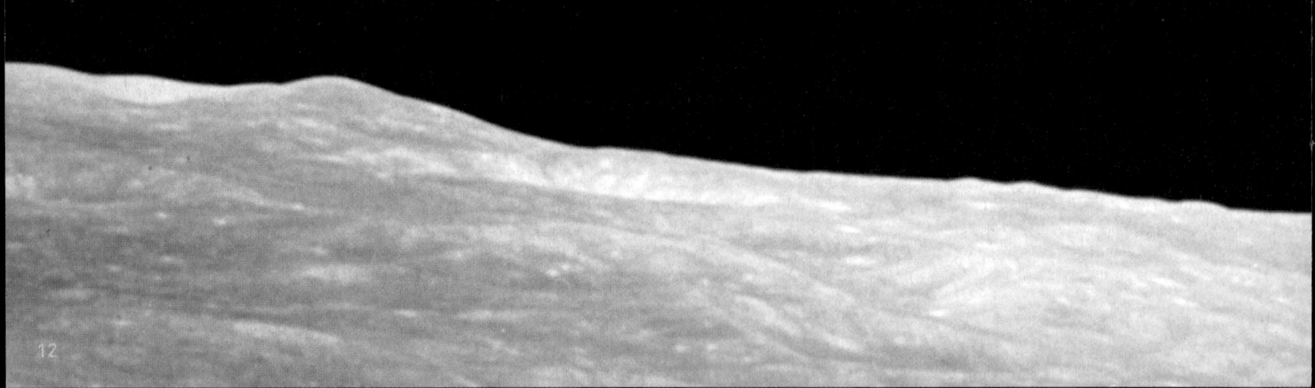

These magical images of Earth were created by a friend of mine, Tom Van Sant. He went through three thousand satellite images taken over a three-year period and carefully selected the ones showing a cloud-free view of the Earth's surface. He then digitally stitched together the images to create a composite view of the planet, in which practically all its surface is clearly visible. (Antarctica is not in view.)

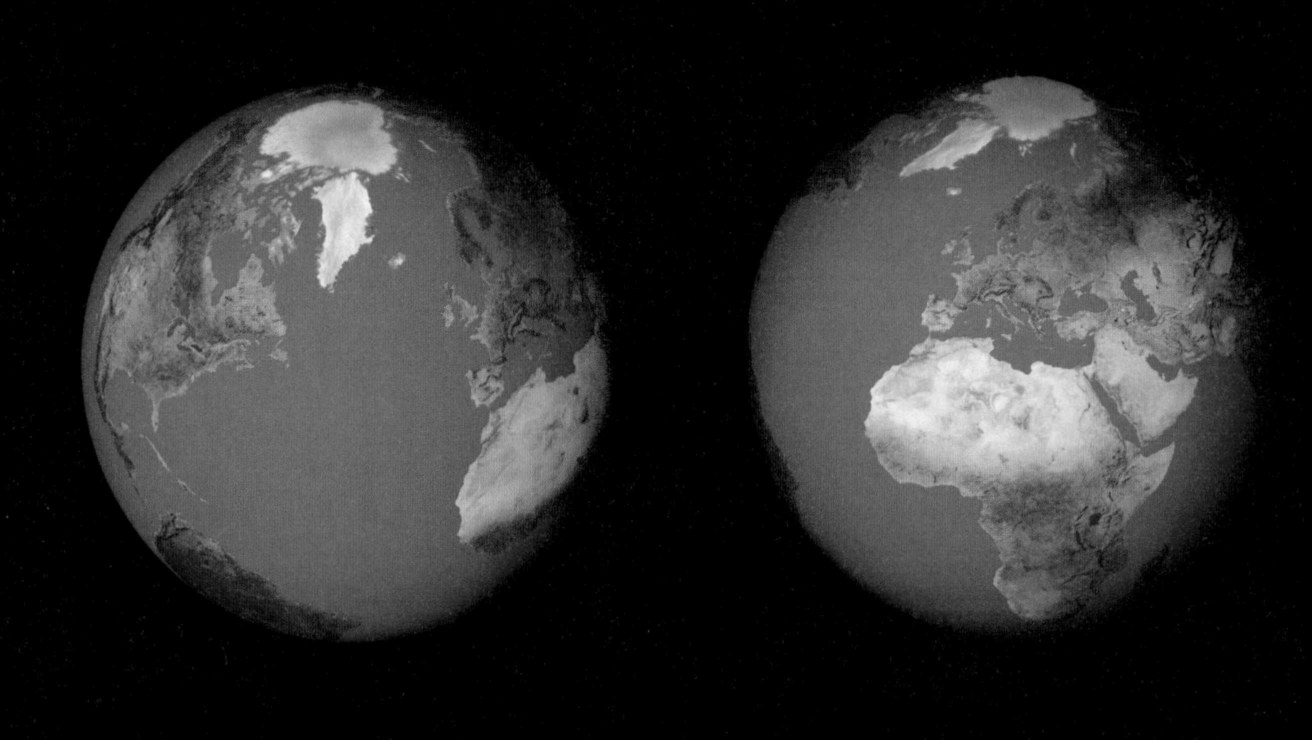

Because the Earth is a globe, the only way to see every part of it in Van Sant's images is to spread them out in a flat picture called a projection. Any projection distorts the shape and size of the continents, especially the area around the North

and South Poles. But the image shown here is based on three thousand of Van Sant's photographs. It has become an icon that is used in many atlases around the world, including *National Geographic*'s.

chapter two

A Silent Alarm

MARK TWAIN ONCE SAID, "It ain't what you don't know that gets you into trouble, it's what you know for sure that just ain't so." His words are especially true in relation to the climate crisis. Many people are convinced, mistakenly, that the Earth is so big, human beings can't do serious damage to it.

Maybe that was true at one time. But not now. There are so many people on Earth (6.5 billion) and technologies have become so powerful that we are capable of causing serious harm to the environment.

What is the most vulnerable part?

The atmosphere surrounding the Earth, because it's such a thin layer.

My friend, the late Carl Sagan, used to say, "If you had a globe covered with a coat of varnish, the thickness of that varnish would be about the same as the thickness of the Earth's atmosphere compared to the Earth itself."

A digitally enhanced photo from space of the sun rising from behind the Earth, 1984

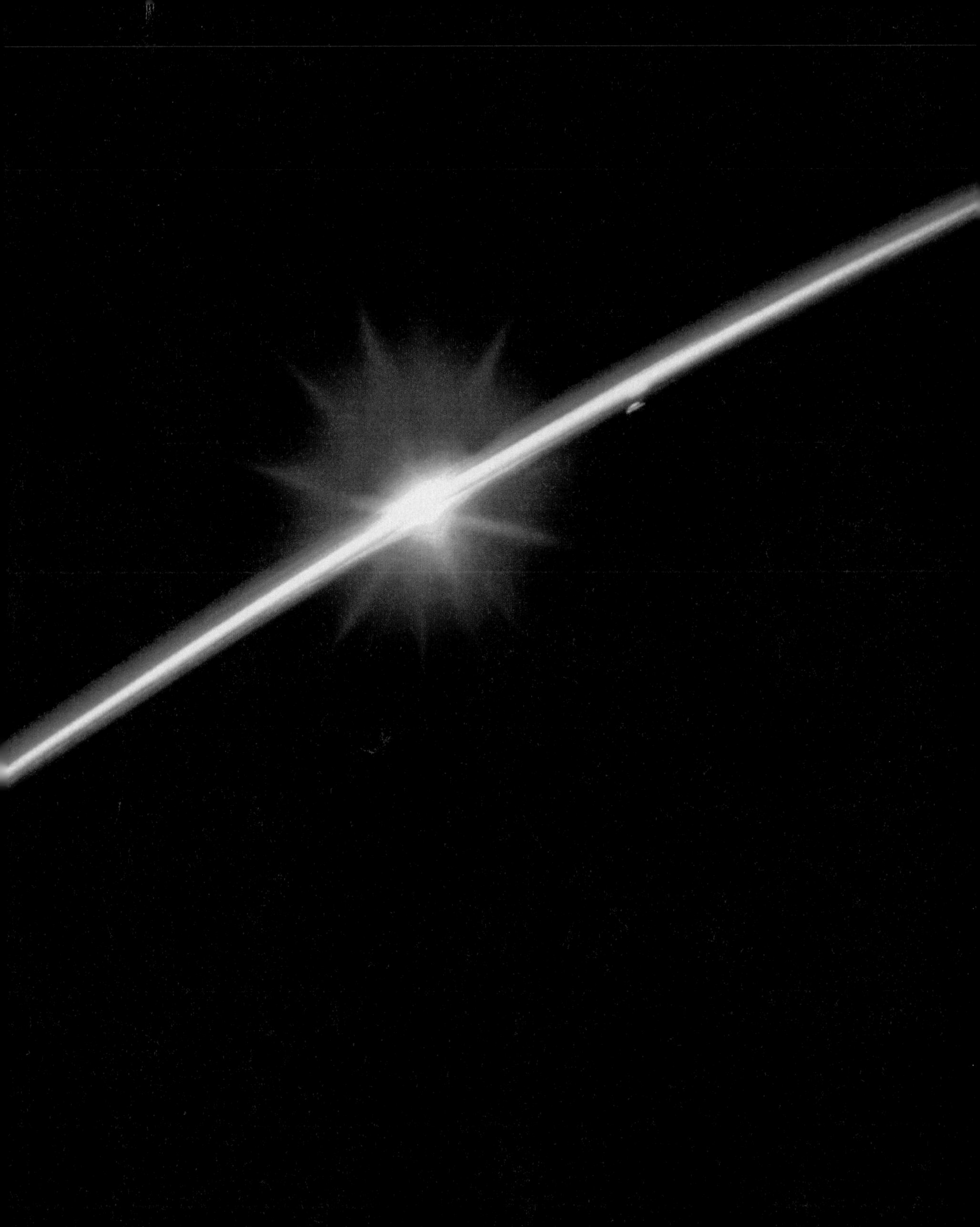

Paper mills like the one shown here are some of the biggest industrial polluters in the United States

The Earth's atmosphere is so thin that we are capable of dramatically changing its composition.

Indeed, we have already dangerously increased the amount of greenhouse gases in the atmosphere.

WHAT EXACTLY ARE GREENHOUSE GASES?

They are gases in our atmosphere that hold in heat, such as carbon dioxide, methane, and nitrous oxide. They maintain an average temperature on Earth of 59°F. Without greenhouse gases, the Earth's surface temperature would drop to around 0°F.

But trouble has arisen because industry, technology, and our modern lifestyle release too much of these greenhouse gases. And that's not good.

Of all the greenhouses gases, carbon dioxide (CO_2) usually gets top billing because it accounts for 80 percent of total greenhouse gas emissions. We release CO_2 in the atmosphere when we burn fossil fuels—oil, natural gas, and coal used in cars, homes, factories, and power plants. Cutting down forests and producing cement also release CO_2.

In a pre-industrial world, just the right amount of the sun's energy was soaked up by greenhouse gases in the atmosphere. It was a wonderfully balanced system and accounts for why Earth is sometimes called the Goldilocks planet—neither too hot like Venus with its thick poisonous atmosphere nor too cold like Mars, which has practically no atmosphere at all.

But when too much of the atmosphere is made up of greenhouse gases, it leads to global warming. These images show how it happens. The sun's energy (the yellow squiggly lines) enters the atmosphere. Some of that energy warms up the earth and its atmosphere and then

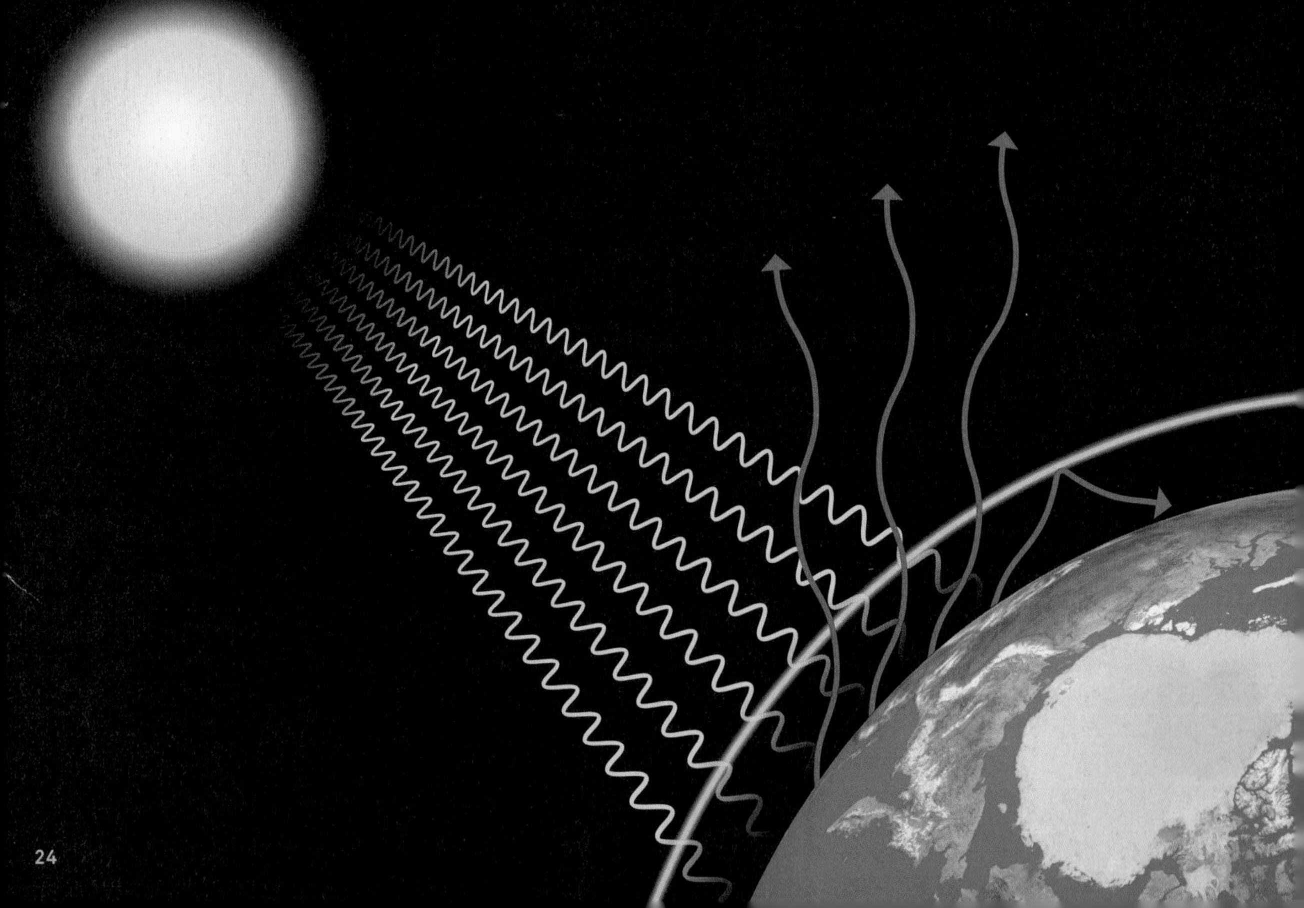

is re-radiated back into space in the form of infrared radiation (the red wavy lines). But greenhouse gases soak up some of the infrared, preventing it from escaping into space.

The problem we now face is that the atmosphere is being filled by huge quantities of human-caused carbon dioxide and other greenhouse gases. This traps a lot of the infrared radiation that would otherwise escape. As a result, the temperature of the Earth's atmosphere and oceans is getting dangerously warmer.

This is what the climate crisis is all about.

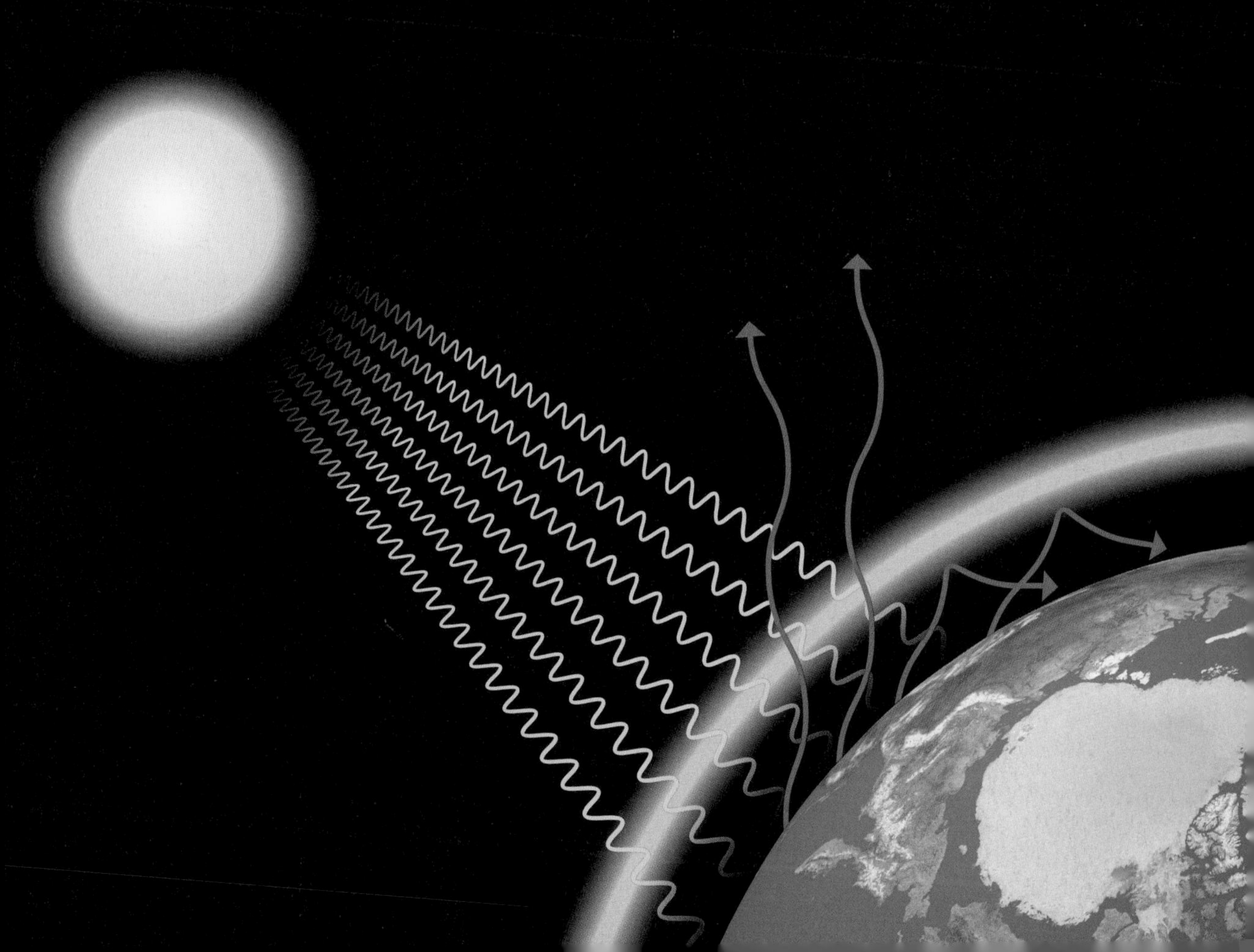

ATMOSPHERIC CO_2 AT MAUNA LOA OBSERVATORY

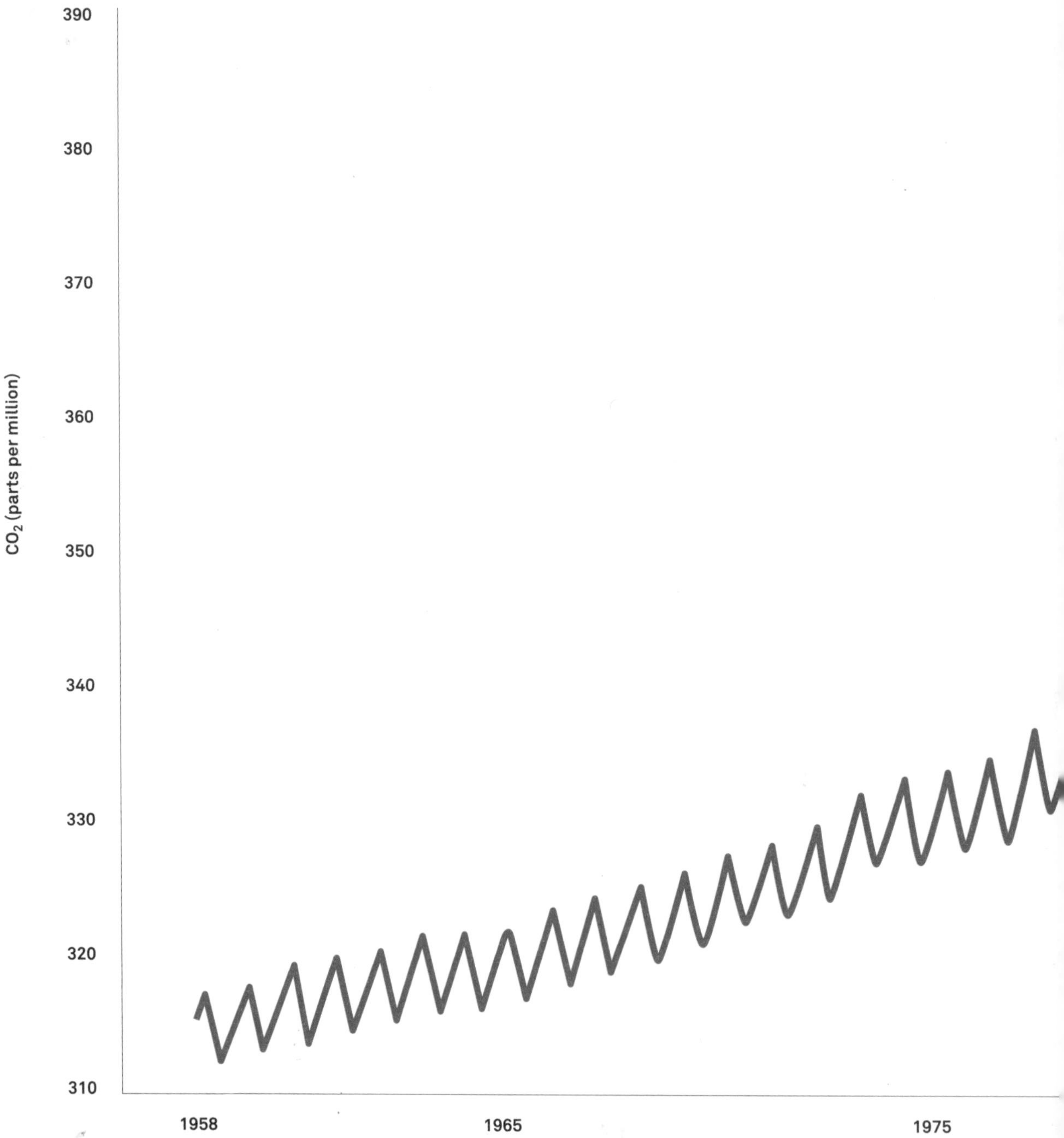

CO_2 (parts per million)

390

380

370

360

350

340

330

320

310

1958

1965

1975

SOURCE: NOAA/SCRIPPS INSTITUTION OF OCEANOGRAPHY

The amount of CO_2 in the atmosphere can be measured. It was my wonderful teacher, Dr. Roger Revelle, who first proposed doing this in 1958. A pattern of steadily increasing CO_2 was visible after the first several years of Revelle's measurements, as seen in this graph. The pattern has continued year by year for almost a half-century. This remarkable and patiently collected daily record now stands as one of the most important series of measurements in the history of science.

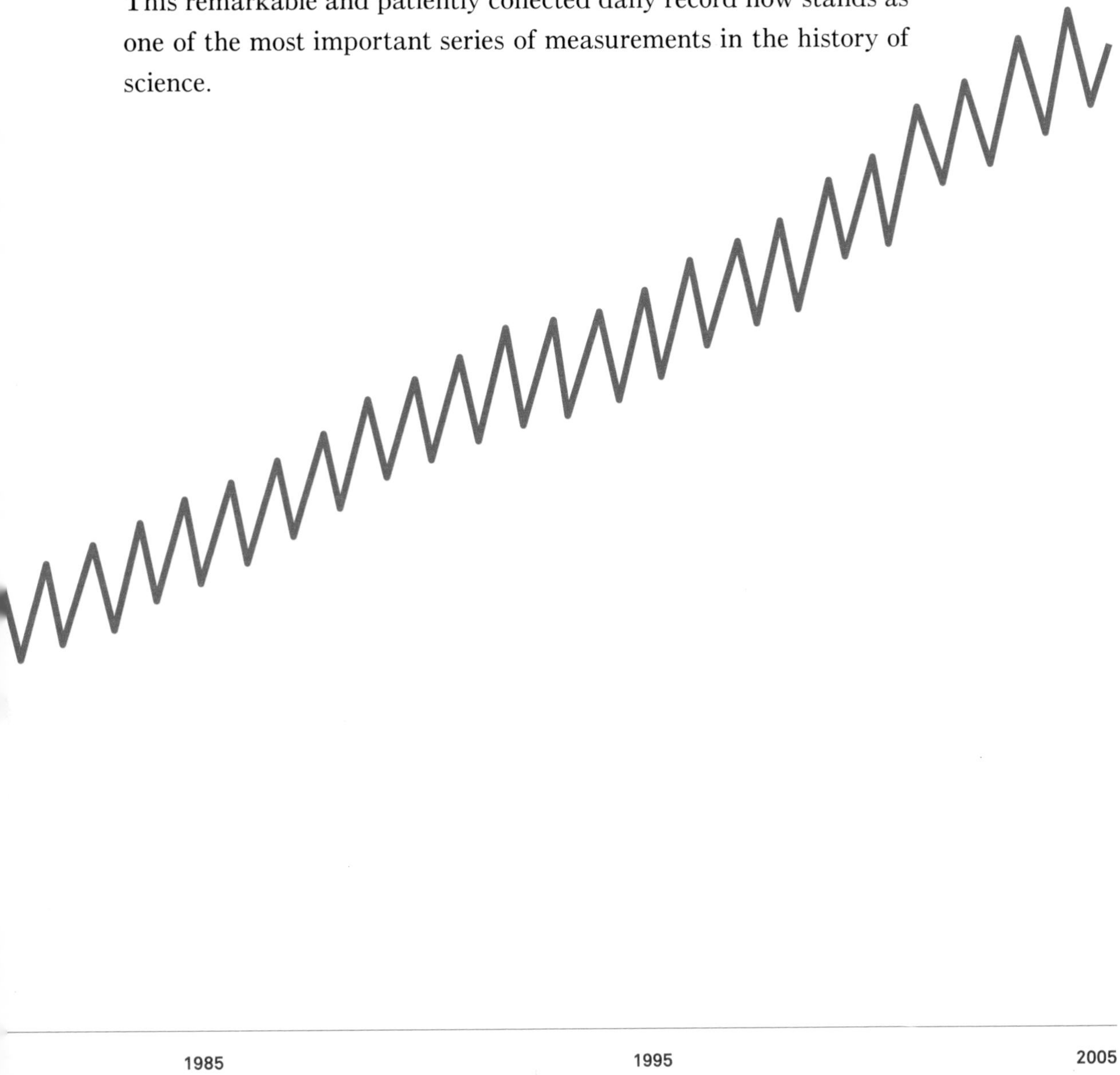

1985 1995 2005

I asked Revelle why the line marking CO_2 concentration goes up sharply and then down once each year. He explained that the vast majority of the Earth's land mass—as illustrated in this picture—

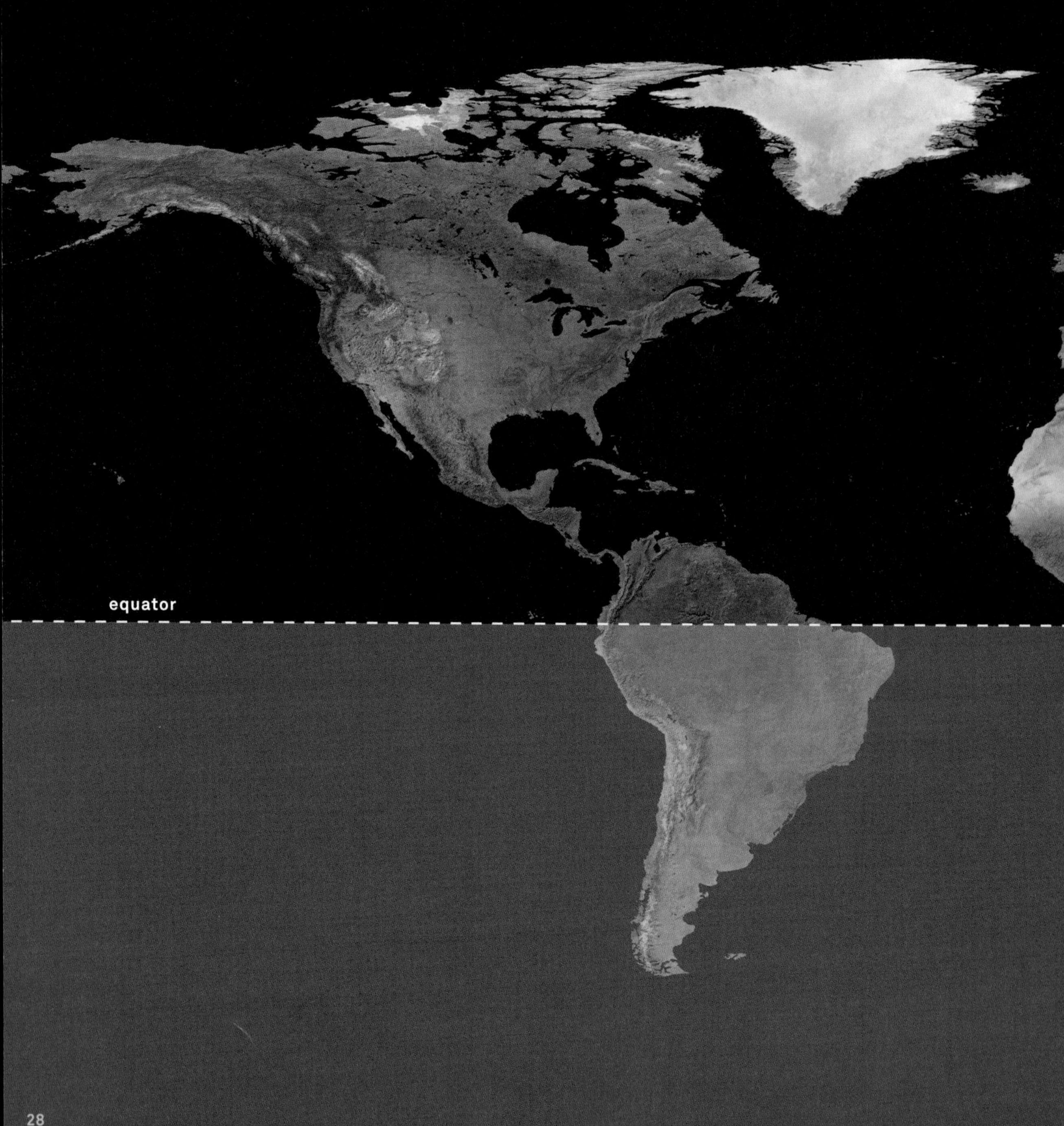

is north of the equator. This means the vast majority of the Earth's vegetation is also north of the equator.

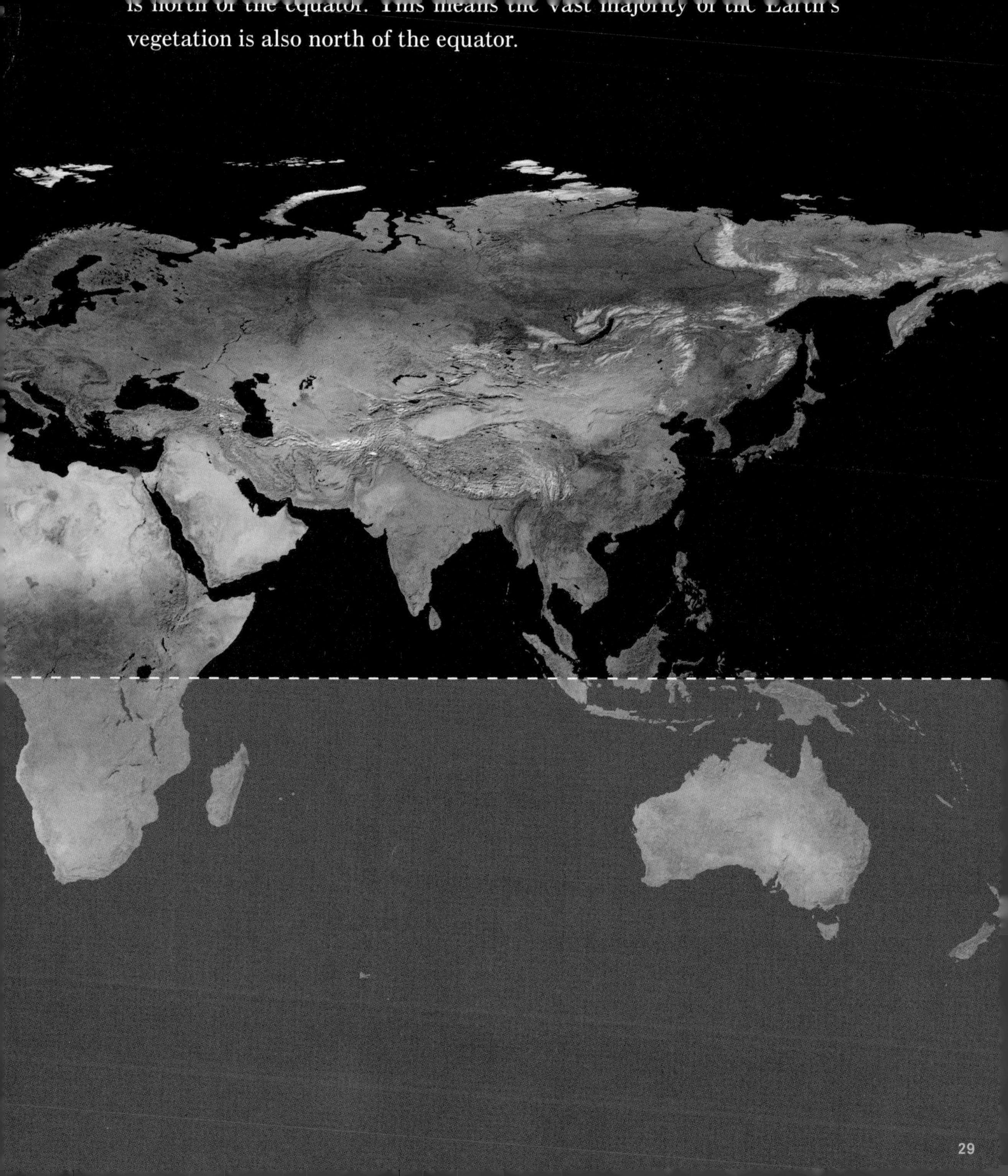

When the Northern Hemisphere is tilted toward the sun during the spring and summer, the leaves come out. They breathe in CO_2, thus decreasing the amount of it worldwide.

However, when the Northern Hemisphere is tilted away from the sun in autumn and winter, the leaves fall and release CO_2. The amount of this gas in the atmosphere goes back up again.

It's as if the entire Earth takes a big breath in and out once each year.

The "inhale" accounts for the yearly dip in CO_2.

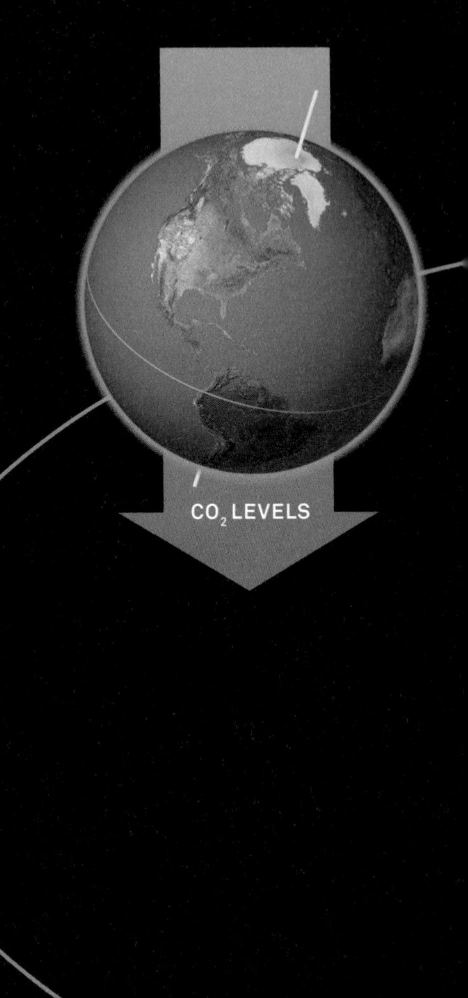

CO_2 LEVELS

chapter three

Cold, Hard Evidence

IT IS EVIDENT in the world around us that very dramatic climate changes are taking place because of global warming.

Mount Kilimanjaro in Africa was once famous for its snow-covered peaks. On the right is what it looked like in 2005. Dr. Lonnie Thompson, the scientist below, predicts that in ten years, the "snows of Kilimanjaro" described by Ernest Hemingway in his novel will have vanished. By the way, he is standing next to what looks like a giant icicle. It was once a glacier!

Mount Kilimanjaro, Tanzania, 1970

Glaciers are huge masses of compacted ice, often very ancient, that flow slowly over land. All over the world they are disappearing.

Our own Glacier National Park will soon need to be renamed "the park formerly known as Glacier."

The glacier on this page was a tourist attraction in the 1930s.

Boulder Glacier, Glacier National Park, Montana, 1932

Now, there's nothing there. I climbed to the top of the biggest glacier in this park with one of my daughters in 1997 and heard from the scientists who accompanied us that within fifteen years all of the glaciers throughout the park will likely be gone.

Boulder Glacier, 1988

Glaciers are vanishing in the Alps, the group of majestic mountains that range over parts of Switzerland, France, Italy, Austria, and Germany.

Here is an old postcard of a Swiss glacier.

Tschierva Glacier, Switzerland, 1910

Here is the same place today.

Tschierva Glacier, 2001

Below is a famous hotel that, a hundred years ago, was perched by another glacier in Switzerland.

Hotel Belvedere, Rhone Glacier, Switzerland, 1906

Here is the same site in 2003. The hotel is still there—but the glacier isn't.

Hotel Belvedere, Rhone Glacier, 2003

Everywhere in the world the story is the same. This photo shows a magnificent glacier on the tip of South America, seventy-five years ago.

Upsala Glacier, Patagonia, Argentina, 1928

The vast expanse of ice is now gone.

Upsala Glacier, 2004

Almost all of the mountain glaciers in the world are now melting, many of them quite rapidly.

Perito Moreno Glacier,
Patagonia, Argentina

The massive Himalayas stretch across Asia and include the fourteen tallest peaks in the world—Mount Everest being the tallest.

The Himalayan glaciers on the Tibetan Plateau have been among the most affected by global warming. The Himalayas contain one hundred times as much ice as the Alps and provide more than half of the drinking water for 40 percent of the world's population, through seven Asian river systems that all originate on the same plateau.

If these glaciers disappear, within the next half-century, 2.6 billion of the world's people may well face a very serious drinking water shortage, unless the world acts boldly and quickly.

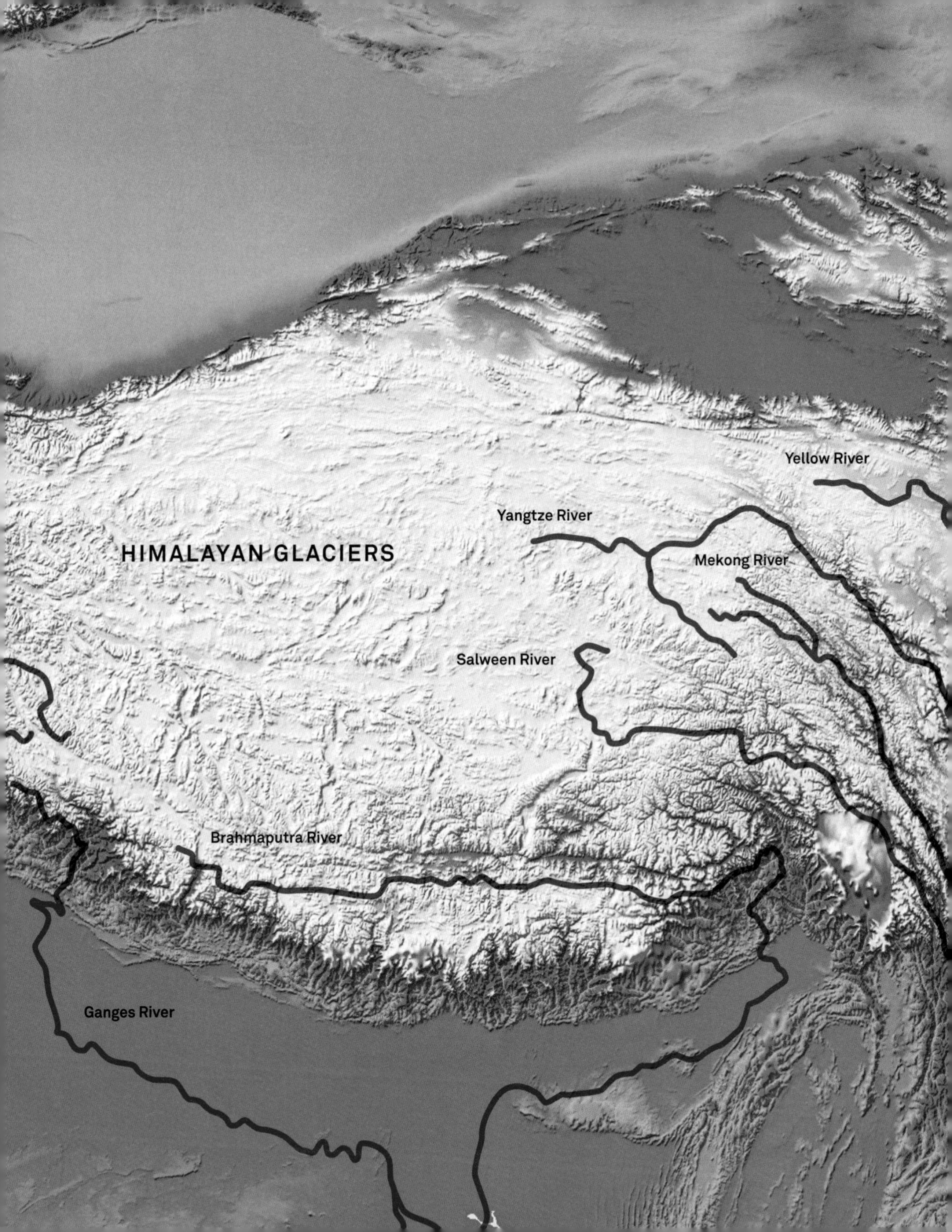
Yellow River
Yangtze River
HIMALAYAN GLACIERS
Mekong River
Salween River
Brahmaputra River
Ganges River

Is there clear evidence that the increase of greenhouse gases is causing glaciers to melt?

YES.

We know from studies of glacial ice.

Glacial ice is deposited in layers, with a new top layer every year. You can see the clear demarcations of the layers in the big photograph to the left. By boring out long cylinders of ice from glaciers all over the world, teams of scientists can measure many things, including the average world temperature each year.

The team can count backward in time year by year, the same way an experienced forester can "read" tree rings, by simply observing the clear lines that separate each year.

Scientists coring through the ice, Huascarán, Peru, 1993

Glaciologist removing an ice core, Antarctica, 1993

Left: Annual layers of ice seen in Quelccaya ice cap, Peru, 1977

This graph of ice-core data shows that in the past couple of centuries, the average yearly temperature has been climbing. It charts a thousand years of temperature averages in the Northern Hemisphere, starting with the year 1000 A.D. and going all the way to 2000 A.D. The line at 0.0 represents the average temperature from 1961 to 1990. Any year with an average temperature below this 0.0 line was colder than average and is shown in blue. Any year with an average temperature above the 0.0 line was warmer than average and is shown in red. Look at all those red years in the recent past.

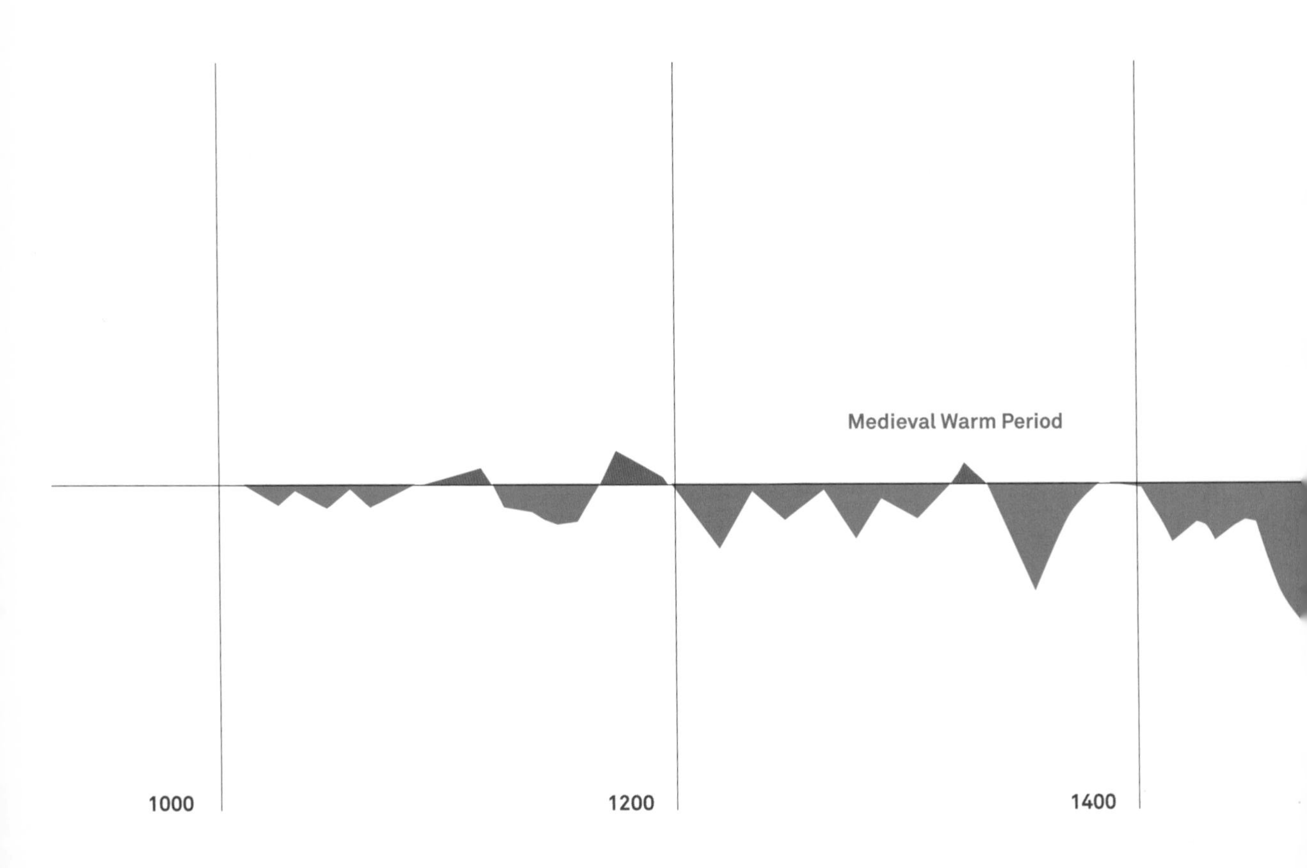

Skeptics often say that global warming is an illusion and that differences in yearly temperatures occur naturally, not from what people are doing to the atmosphere. To support their view they frequently point to the Medieval Warm Period—the little blip of red occurring at around 1350 A.D.

But this spike is tiny compared to the enormous increases in temperature during the past fifty years—at the far right of the chart.

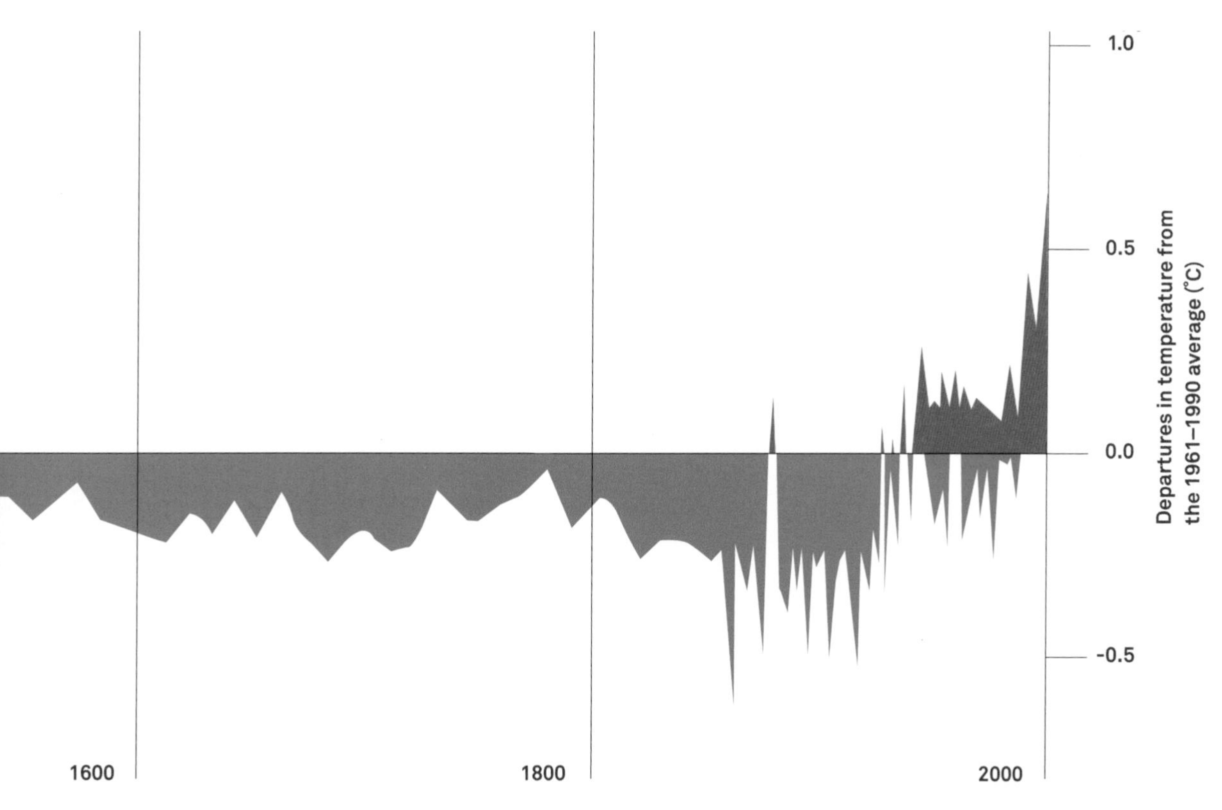

Were the rising temperatures caused by greenhouse gases?

To check, scientist Lonnie Thompson took his team of experts to the tops of glaciers all over the world and drilled down into the ice. Then they examined the tiny air bubbles trapped in the ice cores to measure how much CO_2 was in the Earth's atmosphere in the past 650,000 years, year by year. In the graph below, the blue line charts CO_2 concentrations over this long period, and the gray line shows the world's average temperatures over the same 650,000 years.

If higher amounts of CO_2 are related to higher temperatures, then the blue line and the gray line will follow the same pattern. And indeed they do.

Now the blue line has jumped higher than in all of those 650,000 years. The sudden increase happened in the last two hundred years, during our industrial era. You can see where experts predict CO_2 levels will be in just 45 years by looking at the top red dot to the right. And that means the world's temperatures will just keep going up and up too.

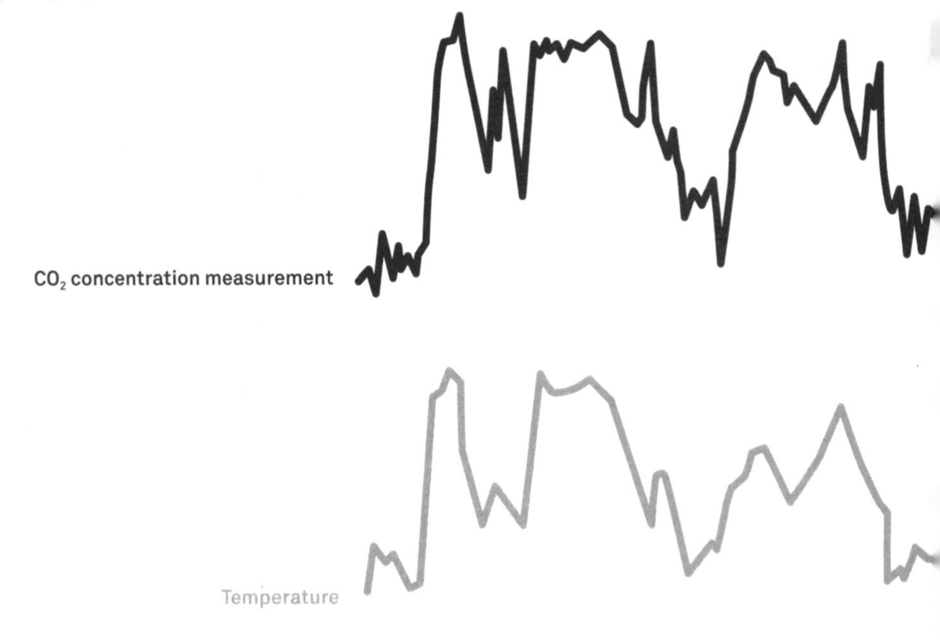

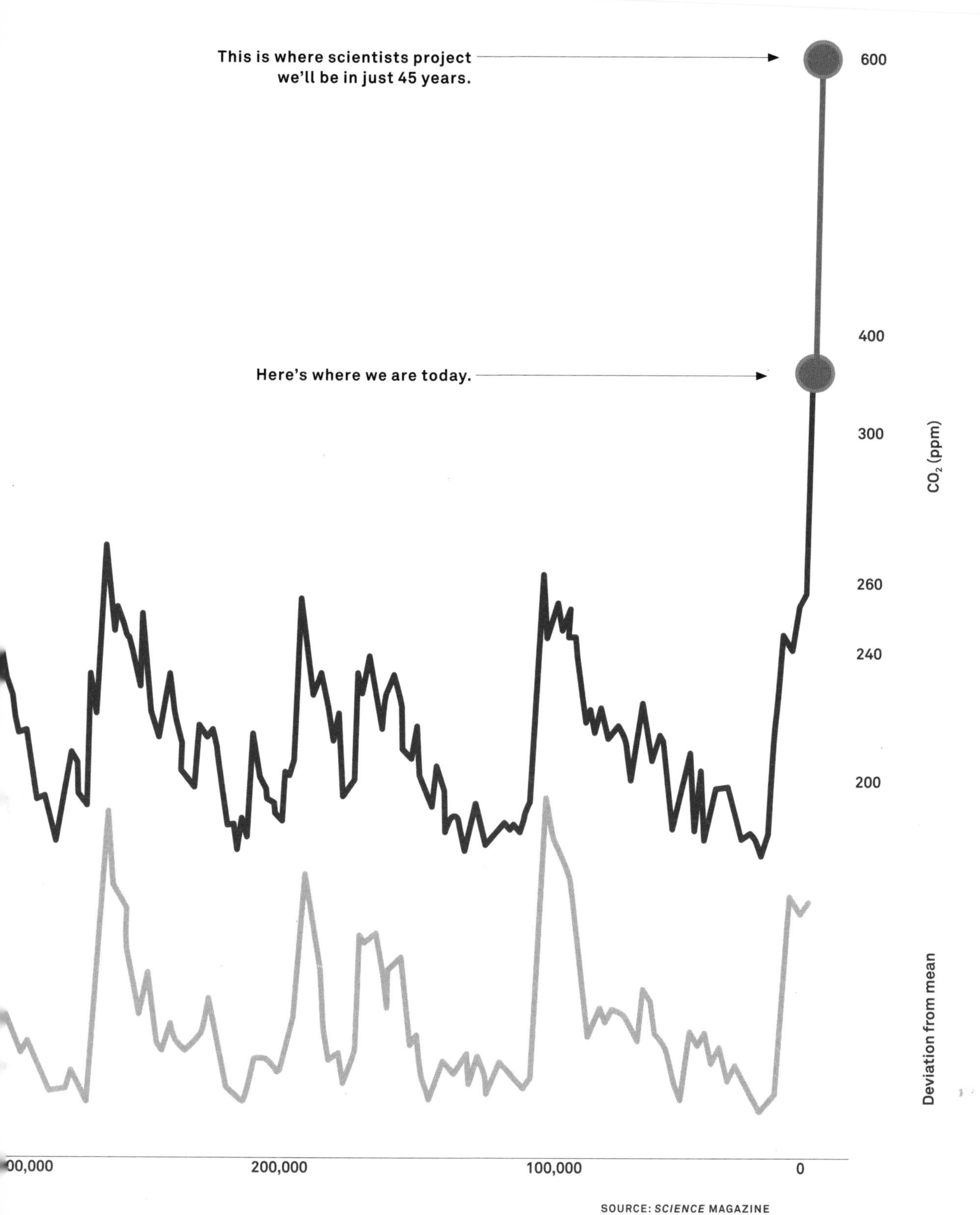

SOURCE: *SCIENCE* MAGAZINE

This graph charts departures from average global temperatures from the time of the Civil War to the present. This is a much narrower timeframe than the last graph, so it offers a "close-up" view of temperature change. The straight line at zero represents the average temperature from 1961 to 1990.

Some recent years in the 1950s through 1970s show dips in average yearly temperature. But the overall trend, shown by the orange line, is one of higher temperatures. And in the most recent years, from the 1980s on, the rate of increase has been accelerating. In fact, of the twenty-one hottest years measured, twenty of them were in the last twenty-five years.

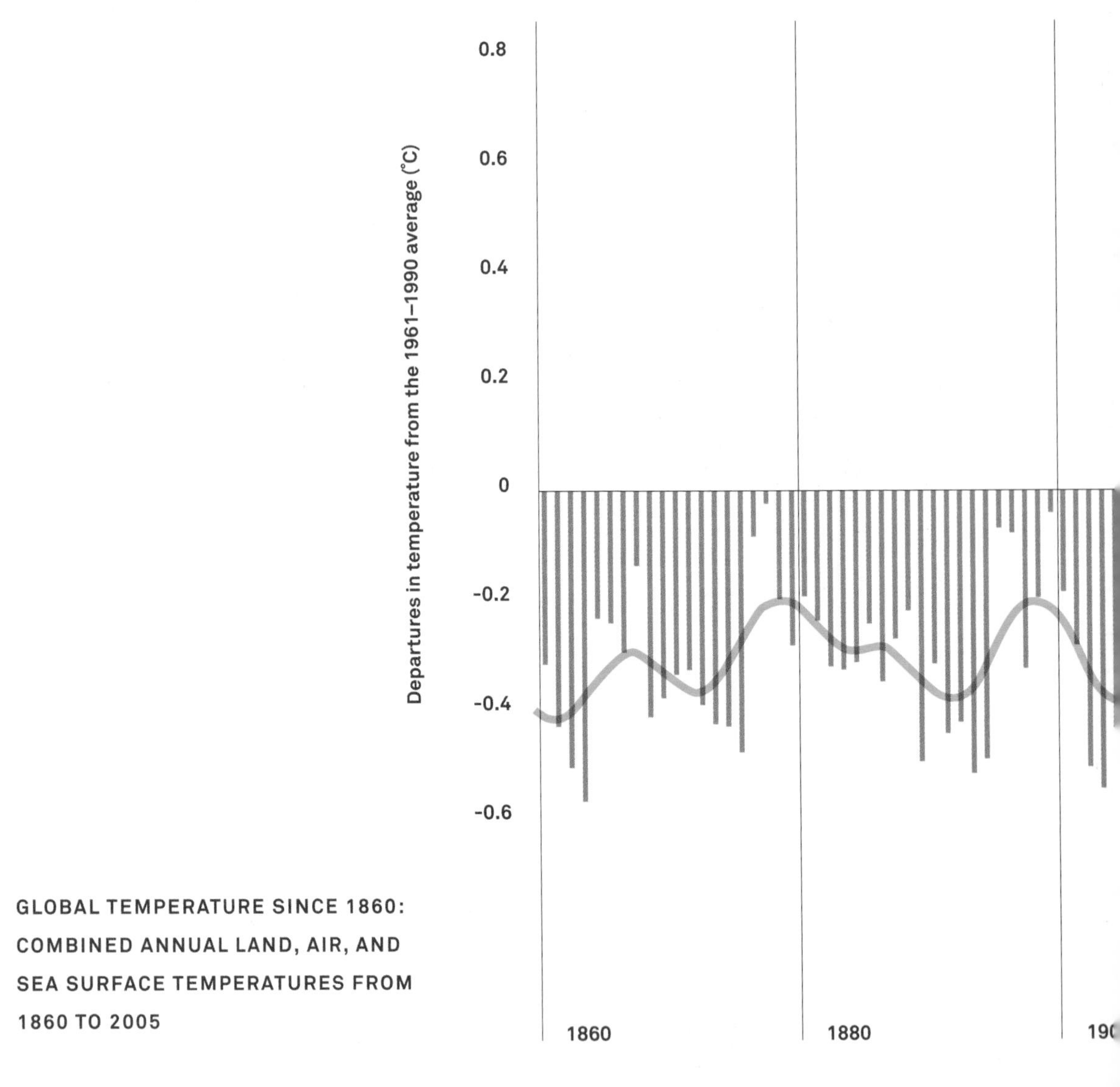

GLOBAL TEMPERATURE SINCE 1860: COMBINED ANNUAL LAND, AIR, AND SEA SURFACE TEMPERATURES FROM 1860 TO 2005

The hottest year recorded during this entire period was 2005.

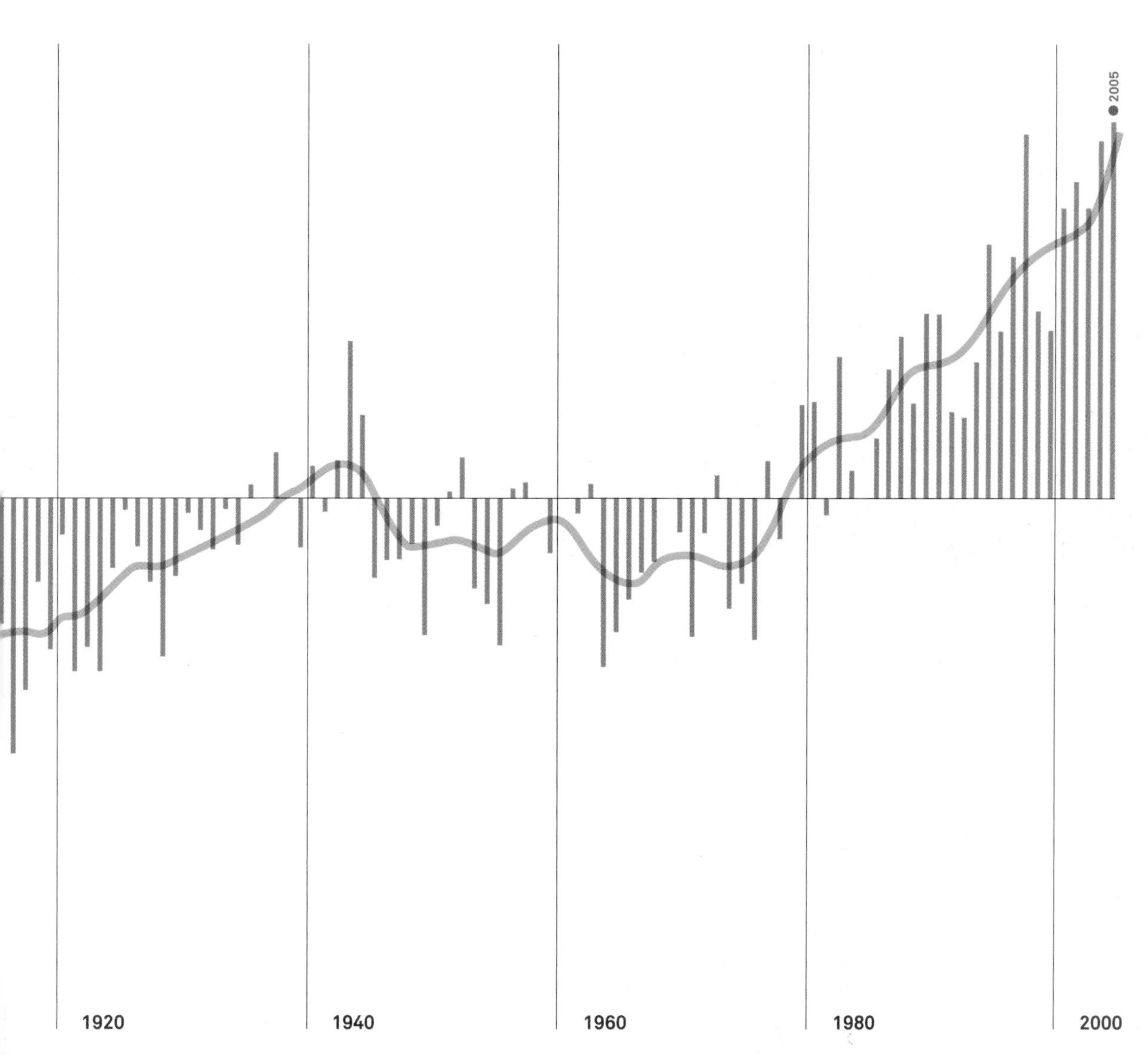

SOURCE: IPCC

We have already begun to see the heatwaves scientists say will become more common if global warming is not addressed. In summer 2003, a massive heatwave in Europe killed 35,000 people.

Munich Zoo during heatwave, Germany 2003

In the summer of 2005 many cities in the American West broke all-time records for high temperatures and for the number of consecutive days with temperatures of 100°F or more. In all, more than two hundred cities and towns in the West set all-time records.

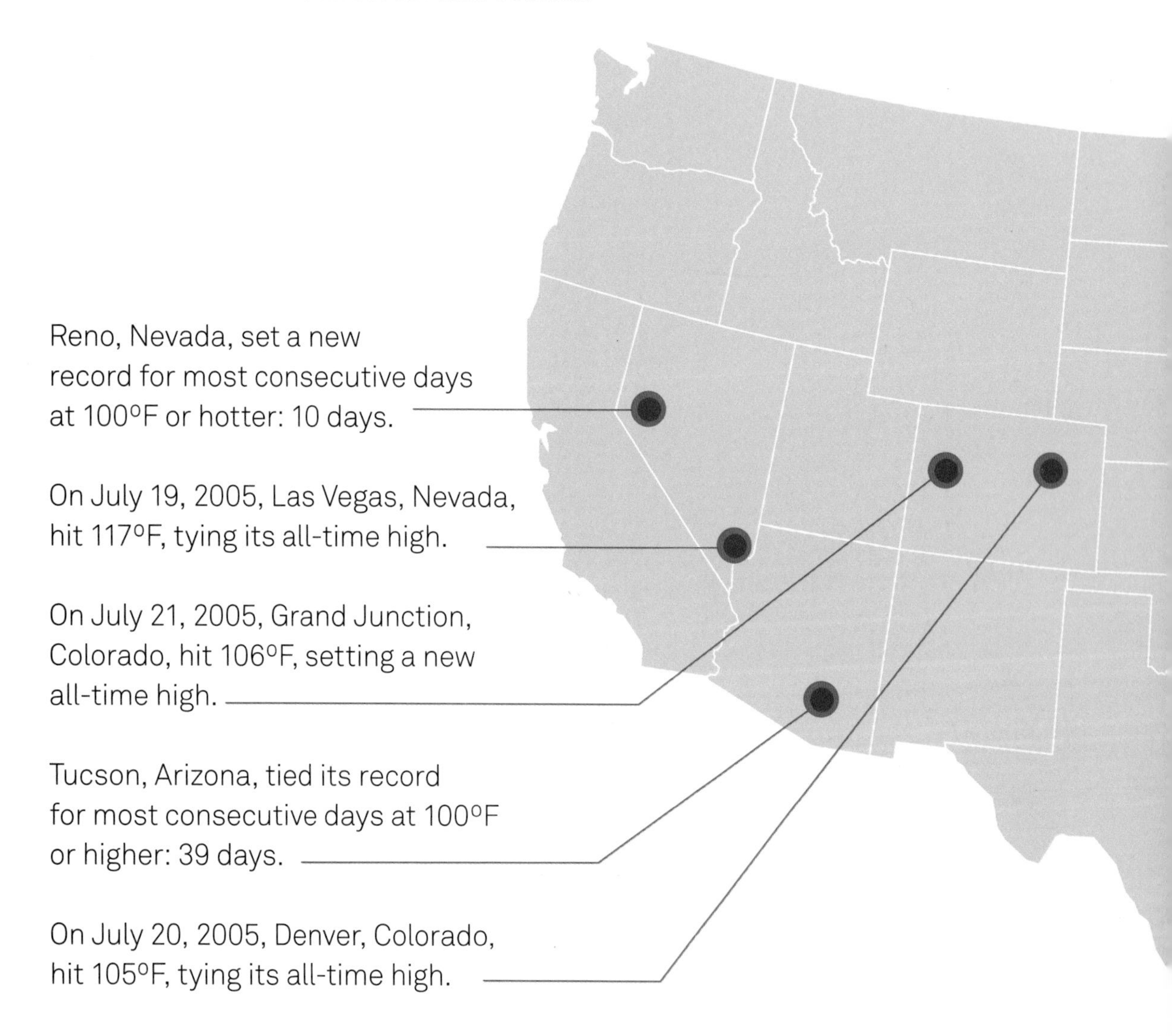

And in the East, a number of cities set daily temperature records, including New Orleans.

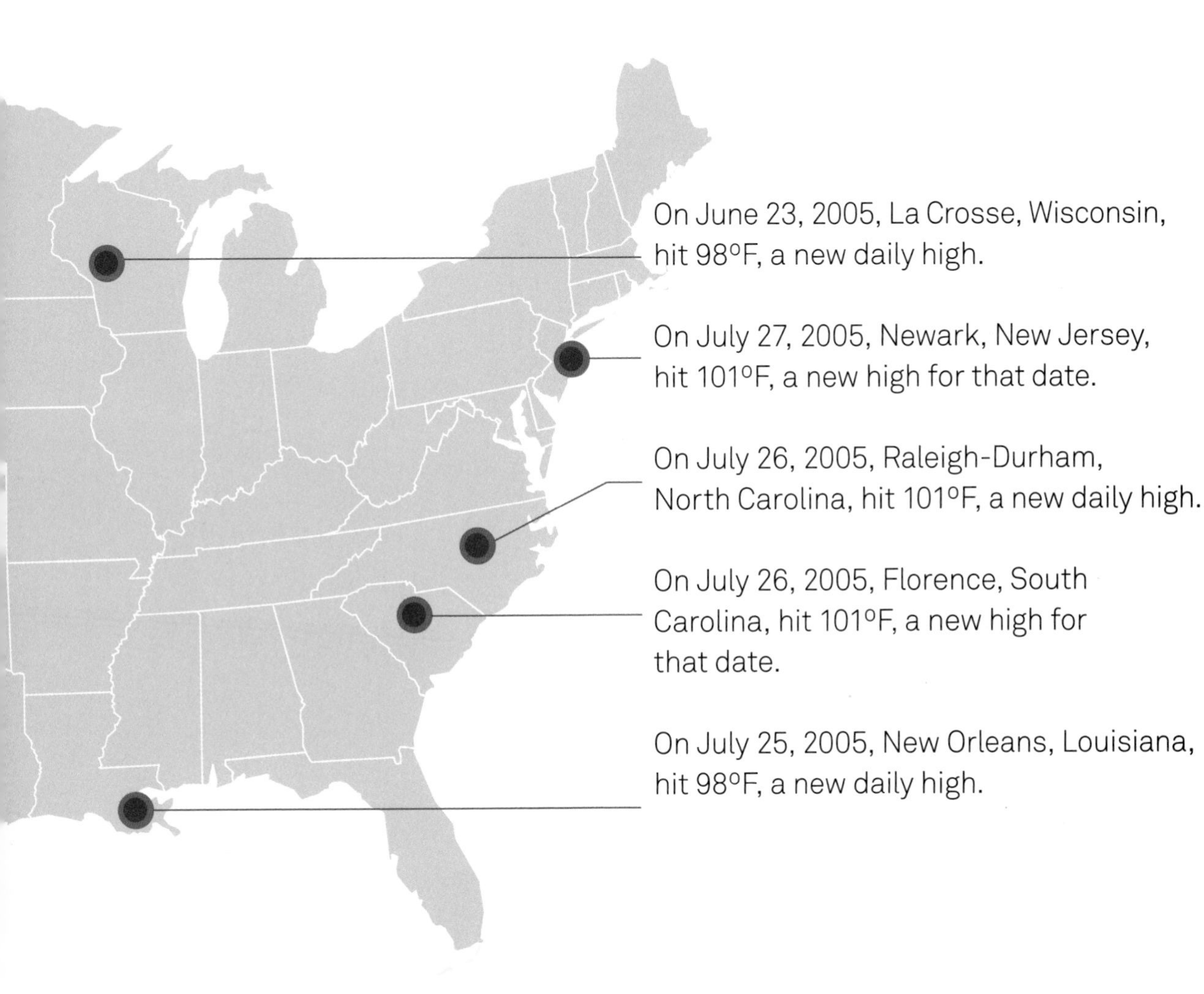

SOURCE: NATIONAL WEATHER SERVICE

Higher temperatures dry out soil, which needs to hold in moisture in order for healthy crops to grow. If, here in the United States, we continue to add CO_2 into the atmosphere at the rate we have been, in less than fifty years, vast areas of our farmland will dry out.

Farmer in drought-ravaged field, Wharton County, Texas, 1998

Because of drier soil and leaves, wildfires are becoming much more common. In addition, warmer air produces more lightning, another cause of fire. The graph below shows the steady increase in major wildfires in North and South America over the last fifty years, decade by decade. The same pattern is found on every other continent . . . except for the icy, treeless expanse of Antarctica.

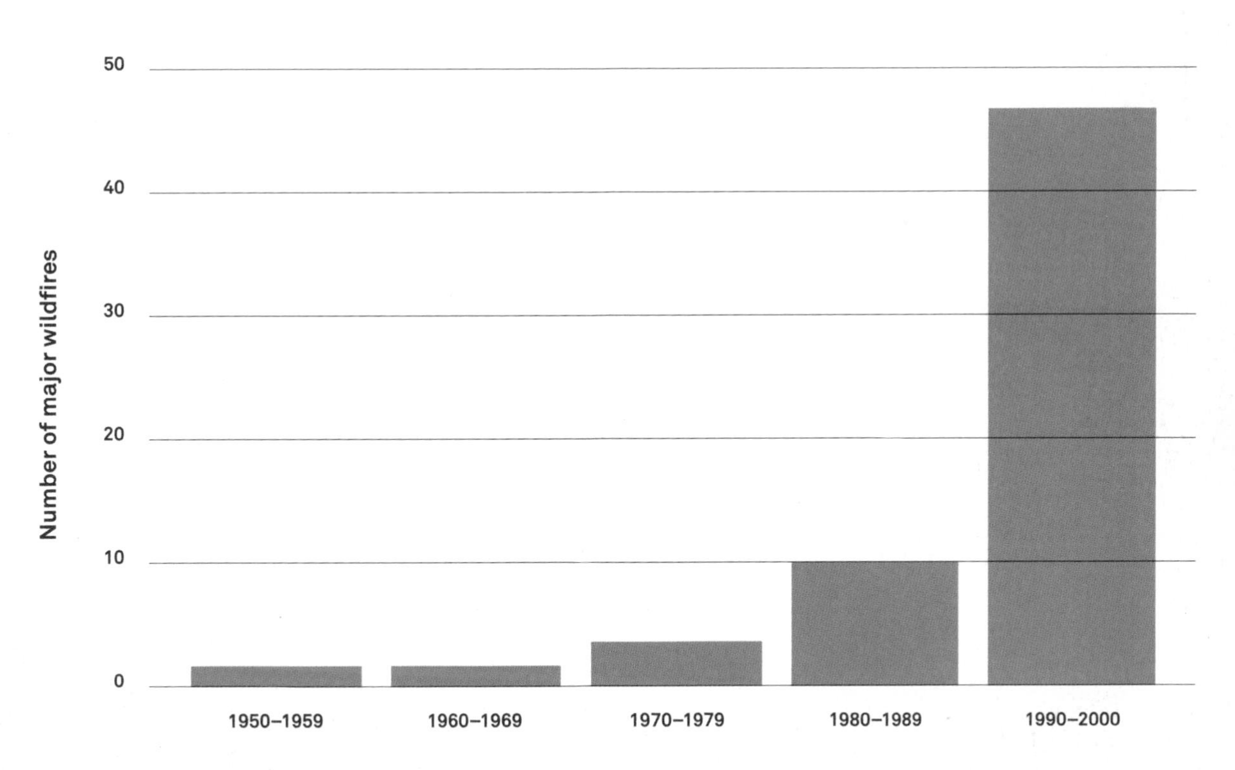

SOURCE: MILLENNIUM ECOSYSTEM ASSESSMENT

chapter four

Hurricane Watch

WARMER WATER near the ocean's surface fuels more frequent, big hurricanes, as a growing number of new studies confirms. (A hurricane is an ocean-based storm with winds of at least 74 miles per hour.) A strong agreement is now emerging that global warming is leading to more frequent, big hurricanes.

There are hurricanes now in parts of the world where they never used to occur. Science textbooks needed to be rewritten in 2004. They had said it was impossible to have hurricanes in the South Atlantic. But that year, for the first time ever, a hurricane hit Brazil.

Hurricane Ivan over the southern United States, September 2004

Two thousand four was also a record storm year in Japan, where typhoons rage during the summer months, reaching winds of 180 miles per hour. (A typhoon is the same kind of storm as a hurricane; storm types are identified by which ocean they originate in.) In 2004 ten typhoons hit Japan. The previous record had been six.

Typhoon Namtheun, off the coast of Japan, July 2004

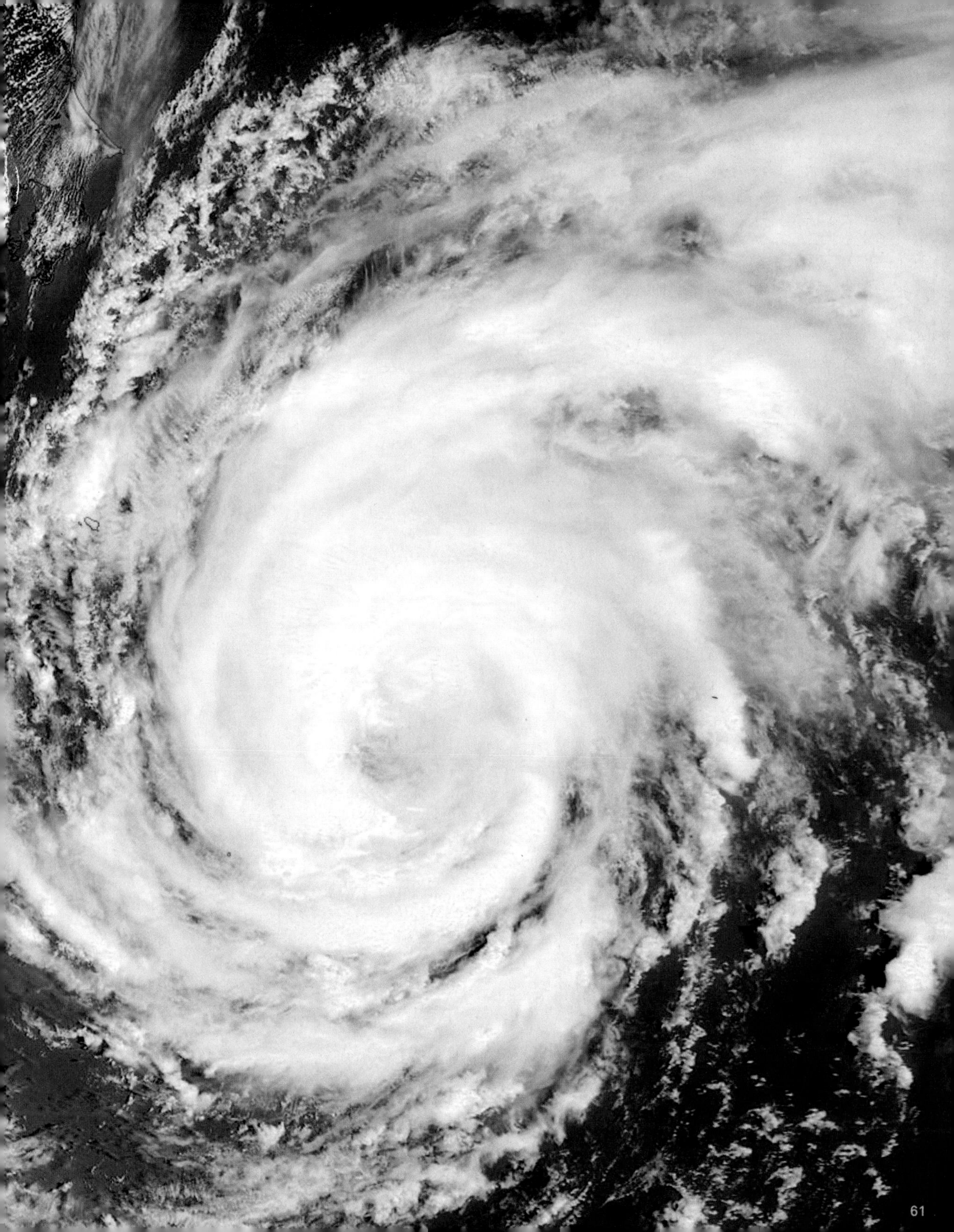

Many tornadoes—inland storms that take shape as huge rotating funnels of wind—are caused by remnants of hurricanes. In 2004, the all-time record for tornadoes in the United States was broken.

Storms are coming with greater force.

There has been a significant increase in the number of the fiercest hurricanes—category 4, with winds between 131 and 155 miles per hour, and category 5, the top category, with winds of more than 155 miles per hour.

In 2005, less than a month before Hurricane Katrina, a study by the Massachusetts Institute of Technology showed that, since the 1970s, storms in both the Atlantic and Pacific have increased in intensity by about 50 percent. That's a staggering fact. Even if we don't live in the path of one of these super storms, all we have to do to witness their destructive powers is turn on the news. The MIT study supports the agreement among scientists that global warming is to blame for the storms' greater strength.

There were so many tropical storms and hurricanes in 2005 that we ran out of names for them. For the first time, the World Meteorological Organization had to start using the letters of the Greek alphabet to name storms that continued into December, well past the normal hurricane season.

Damage caused by Hurricane Emily, La Pesca, Mexico, July 2005

Arlene

Bret

Cindy

Gert

Harvey

Irene

Jose

Nate

Ophelia

Philippe

Rita

Wilma

Alpha

Beta

Gamma

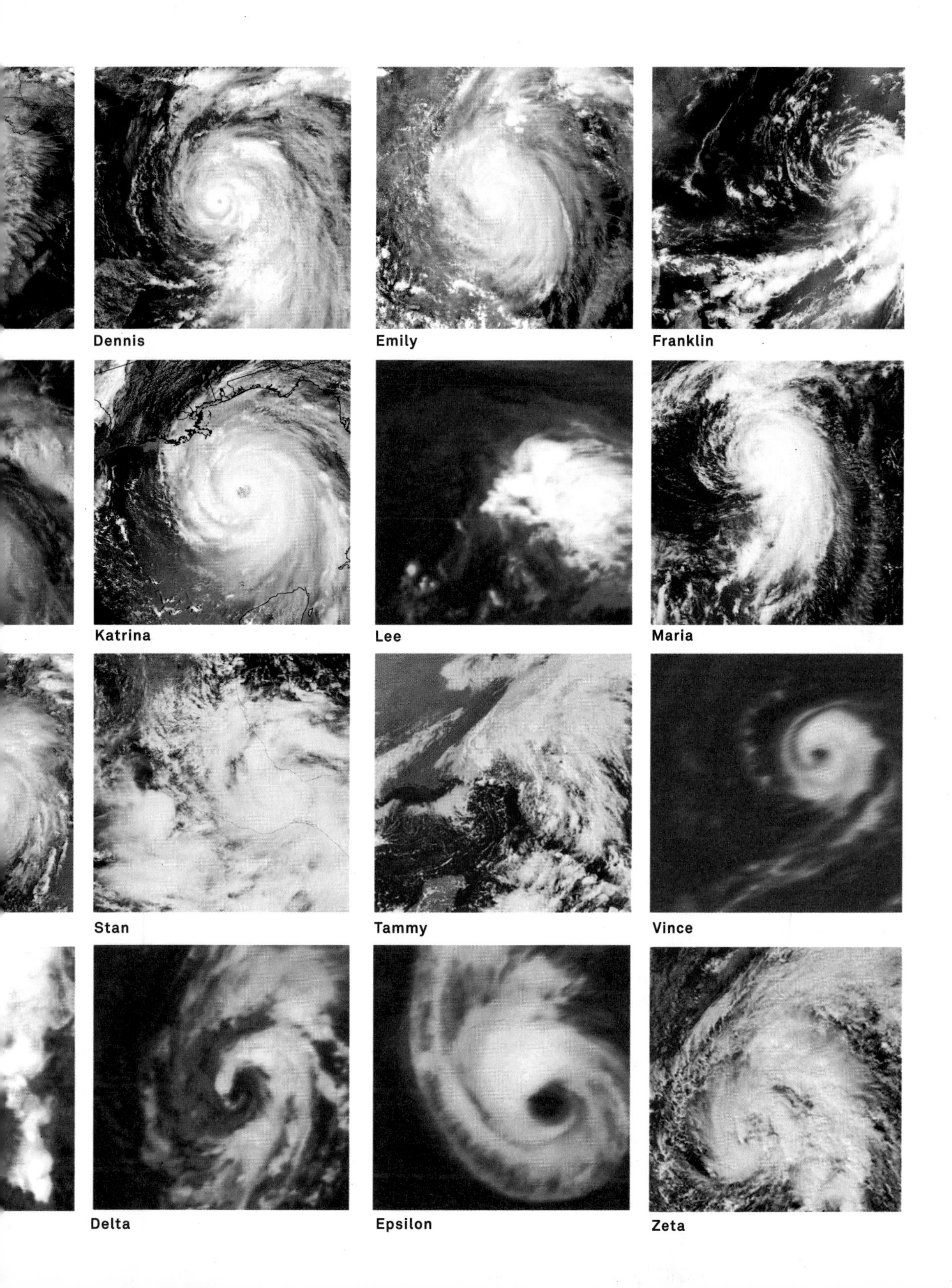

Dennis Emily Franklin

Katrina Lee Maria

Stan Tammy Vince

Delta Epsilon Zeta

The most dramatic storm hit at the end of August 2005—Hurricane Katrina. In New Orleans, the levee system built to keep out flooding waters failed, leaving the city submerged and devastated.

When it first smashed into the coast of Florida, Katrina was only a category 1 storm. Even so, it killed a dozen people and caused billions of dollars in damage.

Then it passed over the unusually warm waters of the Gulf of Mexico. At that point it increased to category 5, one of the most massive storms ever, with winds reaching 175 miles an hour. Although it had decreased to category 3 by the time it reached New Orleans, Katrina caused horrendous destruction.

There are no words to describe it.

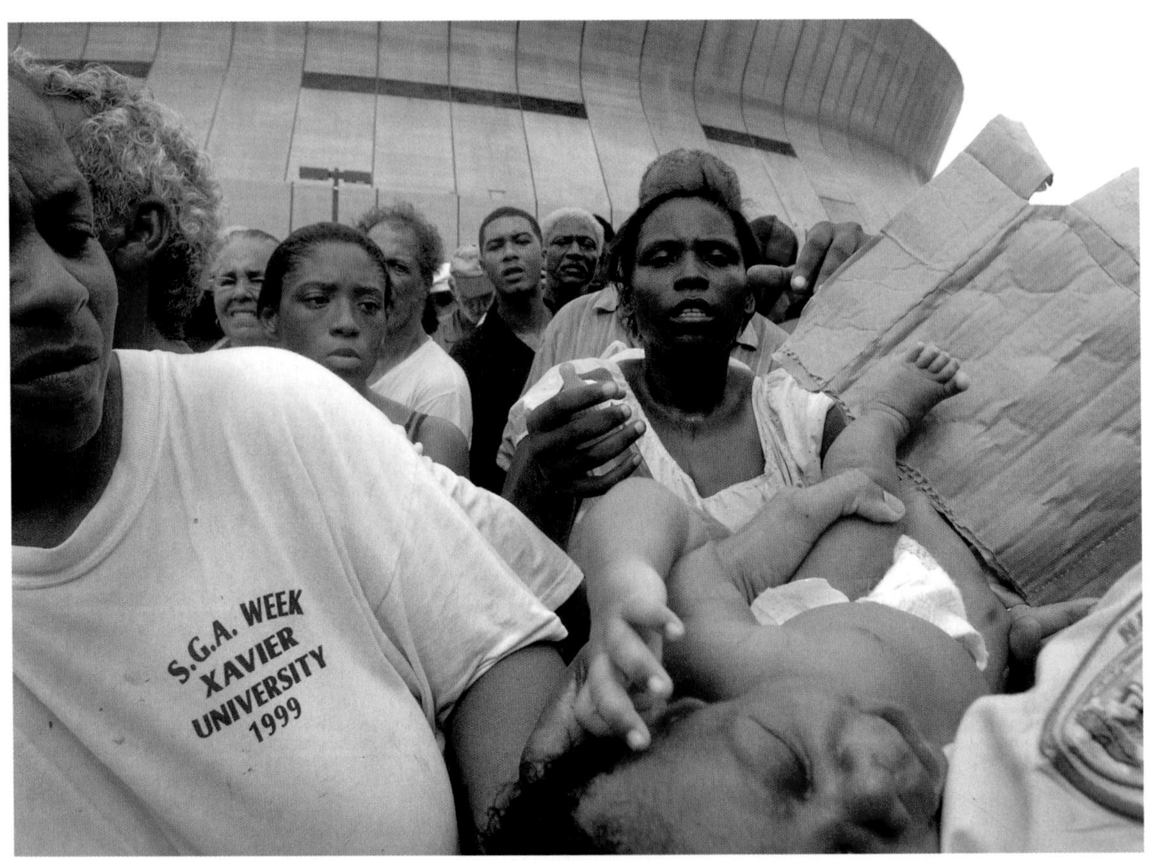

Evacuees of Hurricane Katrina outside the Louisiana Superdome, New Orleans, September 2005

Refugees from Hurricane Katrina in the Astrodome, Houston, Texas, September 2005

In this photo, taken six months later, in February 2006, cleanup in the Lower Ninth Ward of New Orleans had scarcely begun

One of America's greatest cities had been literally drowned. More than a year later, the recovery is only in the beginning stages.

chapter five

Extremely Wet, Extremely Dry

IN THE SUMMER of 2005, while the U.S. was being battered by hurricanes, Europe was experiencing a disastrous series of floods.

Warmer water increases the moisture content of storms, and warmer air holds more moisture. So when there's a downpour, more of it falls as a big, one-time rainfall or snowfall. This, in part, accounts for the increased number of floods, decade by decade, on every continent.

Look at the graph below. From 1950 to 1960, there were fewer than fifty floods in all of Asia (the red column at the far left). Fast forward to the 1990s (the red column at the far right)—there were 325 floods in Asia in the same period of time!

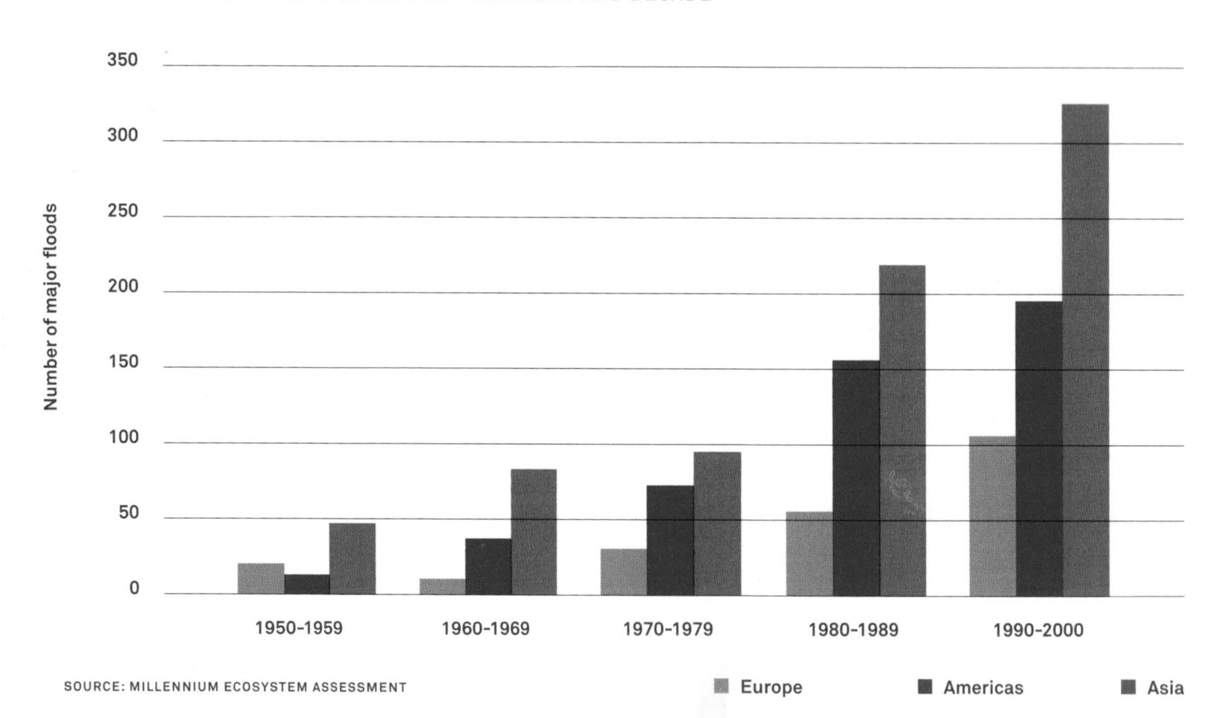

Flooded Schweizerhof Quay, Lucerne, Switzerland, August 2005

Commuters after torrential rains,
Mumbai, India, July 2005

Drought in Anhui Province, China, June 2005

In July 2005, Mumbai, India, received 37 inches of rain in twenty-four hours. It was, by far, the largest downpour that any city in India has ever received in one day. Flood waters rose to seven feet. The death toll in western India reached one thousand. To the left you see rush hour the next day, as thousands of people waded to work.

Global warming relocates rainfall so that certain places experience unusual dry spells. In China in 2005, there was flooding in one province while a neighboring province experienced drought conditions.

Globally, there have been dramatic changes in the amount of precipitation over the last century. On this map, blue dots mark areas all over the world with increased precipitation. The larger the blue dot, the larger the increase. The orange dots show places with decreased precipitation. Again, the larger the dot, the greater the change.

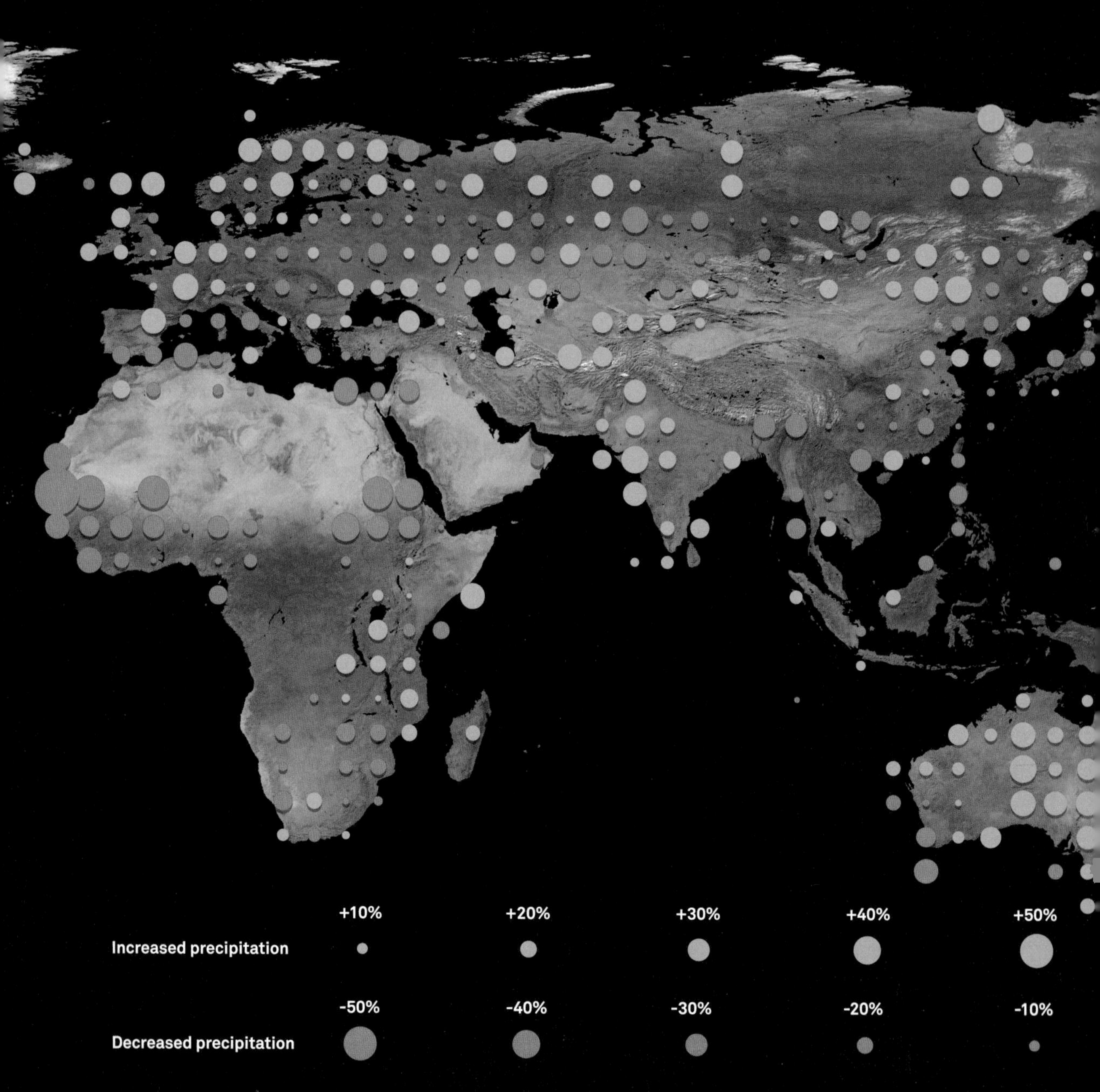

A large change either way can cause devastation to the people of the area.

Look at the places in western Africa on the edges of the Sahara Desert. Always dry, they are now getting as little as half the rainfall they used to. Without rain, crops cannot grow and people starve.

SOURCE: IPCC

The Disappearance of Lake Chad

In central Africa there used to be a huge lake. Forty years ago, Lake Chad was the sixth largest lake in the world, as big as Lake Erie. People in the countries bordering the lake (Chad, Nigeria, Cameroon, and Niger) depended on its waters for crop irrigation, fishing, livestock, and drinking water. Because of declining rainfall and increased human use, Lake Chad has shrunk to one-twentieth its original size.

LAKE CHAD, AFRICA

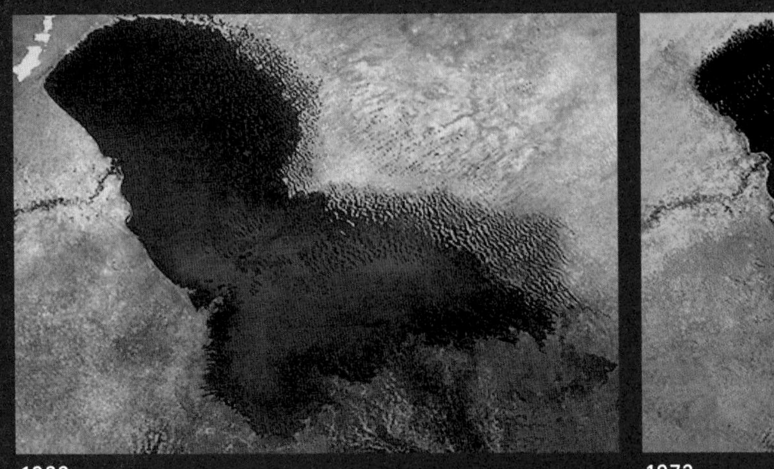

1963

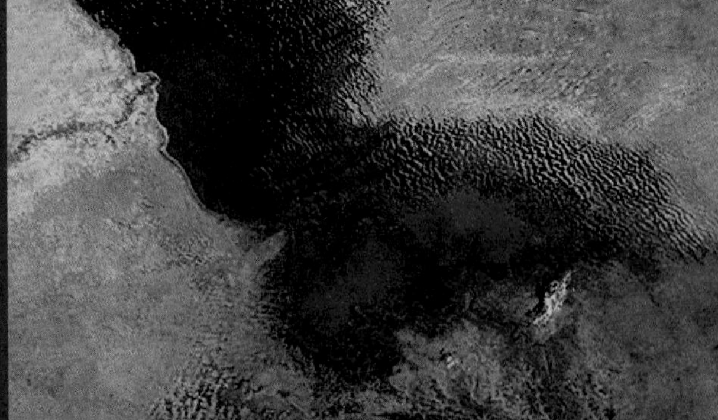

1973

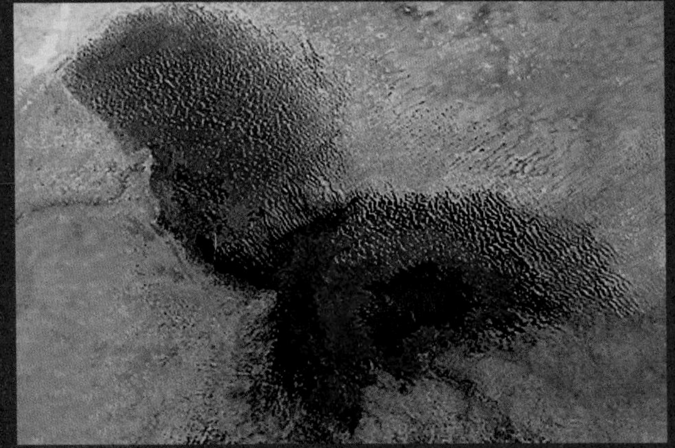

1987

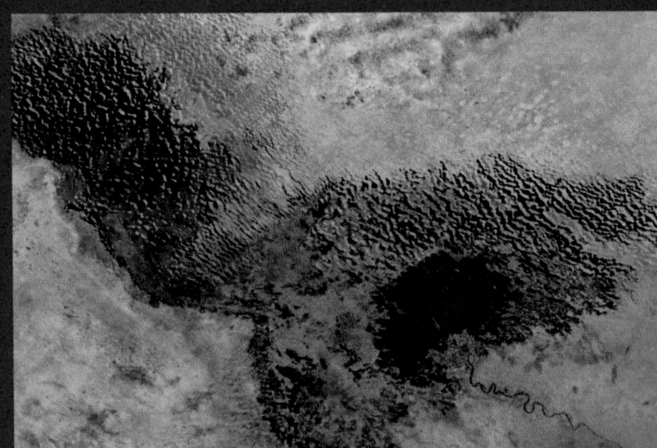

2001

The lake's disappearance has ignited battles between people of neighboring countries where relations were already tense. When Nigerian fishermen followed the receding water into Cameroon, military fighting broke out. Once farmers began to till the former lake bottom, battles began over who actually had the rights to the property.

In Africa, most people still rely literally on the fruits of their labor; they live on the food they grow. When crops fail, things fall apart. In Malawi five million people faced starvation in 2005 when farmers planted crops on schedule but rains failed to come. Five million people! That's over a million more than the entire population of Los Angeles.

A Sudanese mother and her child in a food dispensary, Kalma, Sout Darfur, 2005

Sometimes, it's implied that Africans have brought this situation upon themselve by mismanaging their water supply. But the more we understand climate chang the more it looks as if the advanced nations of the world that have created manmad global warming may be the real cause.

The United States emits about a quarter of the world's greenhouse gases, while the entire continent of Africa is responsible for only about 5 percent. Just as we cannot actually see the greenhouse gases, we do not see their impact on place so far away. But we helped create the suffering ir Africa, and we have a moral obligation to fix it.

chapter six

The Ends of the Earth: The North Pole

THERE ARE TWO places on Earth that are especially vulnerable to global warming—the Arctic and Antarctica. Changes in both areas are happening sooner and more dramatically than anywhere else. Although they may look a lot alike—nothing but snow and ice . . . and more snow and ice—in reality these places are very different. The Arctic is actually ocean surrounded by land while Antarctica is land surrounded by ocean.

SOURCE: COMPOSITE GAZETTEER OF ANTARCTICA, © HER MAJESTY THE QUEEN IN RIGHT OF CANADA, NATURAL RESOURCES OF CANADA, 2001

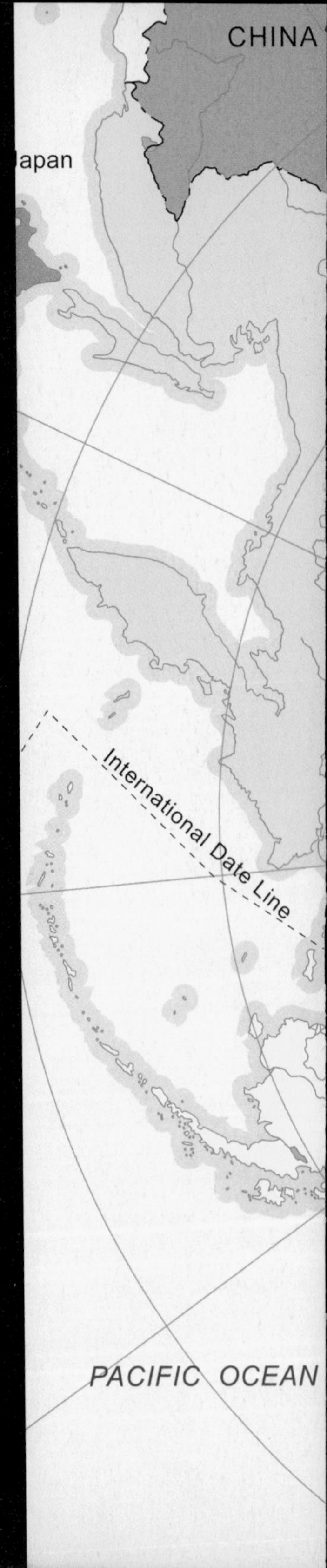

RUSSIA
Ukraine
MD
Belarus
EE
LV
LT
Poland
Finland
RU
Sweden
RU
NO
Norway
DK
NL
United Kingdom
ARCTIC OCEAN
North Pole
International Meridian
DK
Iceland
GREENLAND
DK
ATLANTIC OCEAN
CANADA
Circle

Amazingly, the Arctic ice cap is less than ten feet thick on average. Below it is ocean. You can think of it almost like a huge floating ice rink. The thinness of the ice makes it especially susceptible to effects from global warming. As the ice melts, temperatures are shooting up faster than at any other place on Earth.

An ice shelf is a thick, massive slab of ice attached to the coastline, floating on the ocean. This photo shows the largest one in the Arctic—the Ward Hunt shelf. In 2002 the Ward Hunt ice shelf cracked in half, to the astonishment of scientists. This had never happened before.

Researchers examine the breakup of the Ward Hunt Ice Shelf in Nunavut, Canada, 2002

The Arctic is a region meant to stay frozen all 365 days of the year. Yet the ice cap is melting quickly, in part because it is so thin, and also because it floats on water. The more it melts, the faster it melts.

The illustrations show how ice reflects most of the sunlight, like a giant mirror. The open sea water, however, absorbs most of the heat from sunlight instead. As the water warms up, it causes more melting at the edges of the ice. The ice cap is shrinking because of global warming. There are now studies showing that if we continue to release greenhouse gases at the rate we have been, the Arctic ice will completely disappear each year during summertime.

This is a dangerous situation for all of us, because the Arctic ice cap plays a very crucial role in cooling the entire planet.

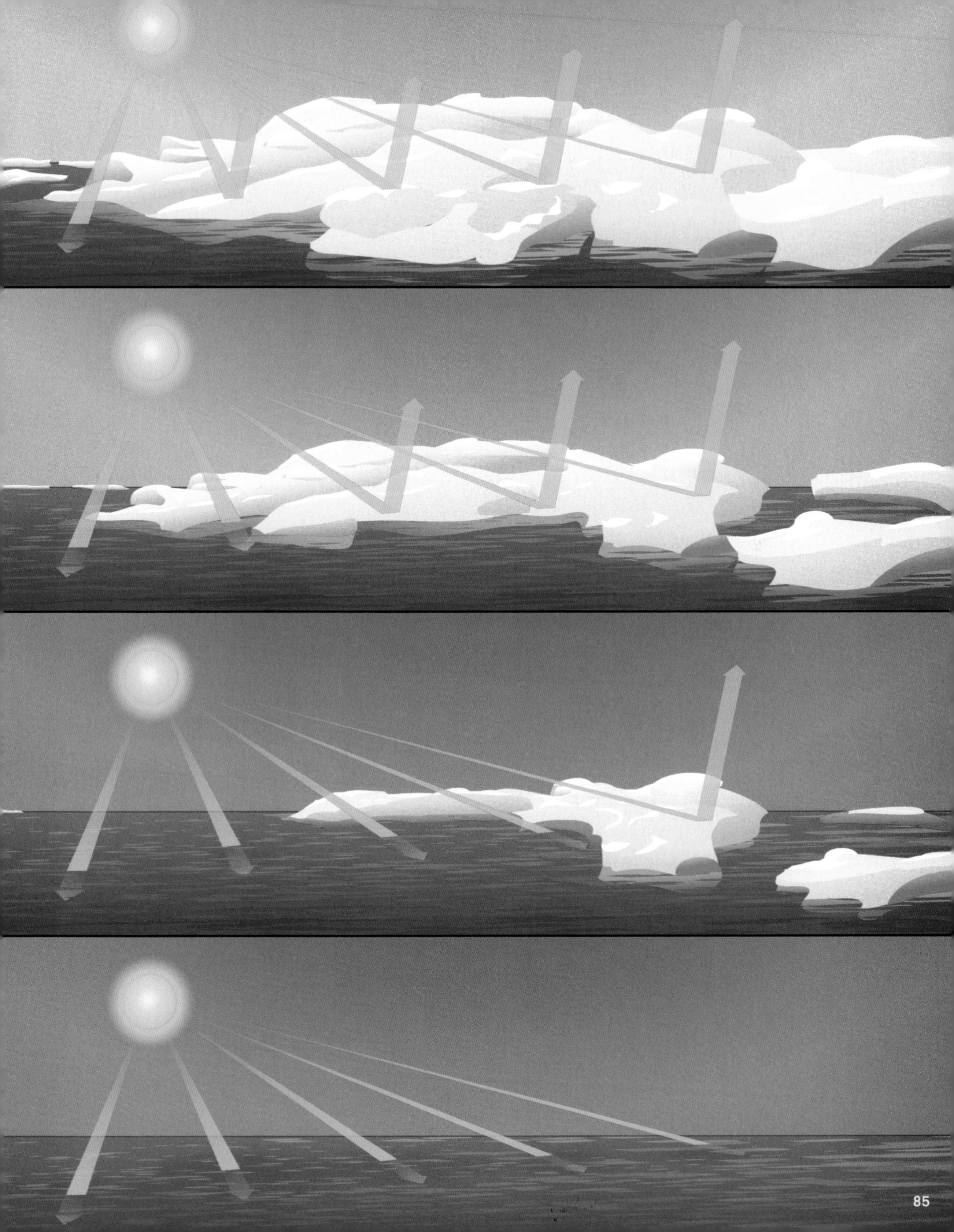

The North Polar ice cap began a fairly rapid retreat in the 1970s. This is bad news for polar bears, which travel from ice floe to ice floe hunting seals. Because so much ice has melted, the bears must now swim longer distances. For the first time, some polar bears are becoming exhausted and drowning before they reach the next ice floe.

Mother polar bear and cub on pack ice, Spitzbergen, Norway, 2002

In Alaska these are called "drunken trees" because they are leaning every which way. These trees had put their roots deep into permafrost (a layer of soil that remains permanently frozen)

decades—sometimes centuries—ago, but now as the permafrost melts, the trees are losing their anchor, causing them to sway in all directions.

Spruce trees north of Fairbanks, Alaska, 2004

A collapsed building in Siberia

The land north of the Arctic Circle is frozen most of the year. However, global warming has begun to thaw large areas of permafrost. That is why these buildings are collapsing. They were built on permafrost that thawed and became too soft to support the structures.

An abandoned house in Alaska

This map shows the areas where frozen tundra—flat, treeless plains found in arctic regions—is thawing. The dot in the center is the North Pole, so you are looking down at a map of the top of the world. The pink areas show where the most severe damage is predicted if thawing continues. The affected part of Siberia represents a vast area of tundra that has been frozen since the last ice age, which ended about 10,000 years ago. According to scientists, seventy billion tons of carbon is locked in this ice. If the tundra continues to thaw and all this carbon is released as methane and CO_2 into the atmosphere, it would be an ecological disaster. The amount escaping would be more than ten times the amount released yearly from manmade sources.

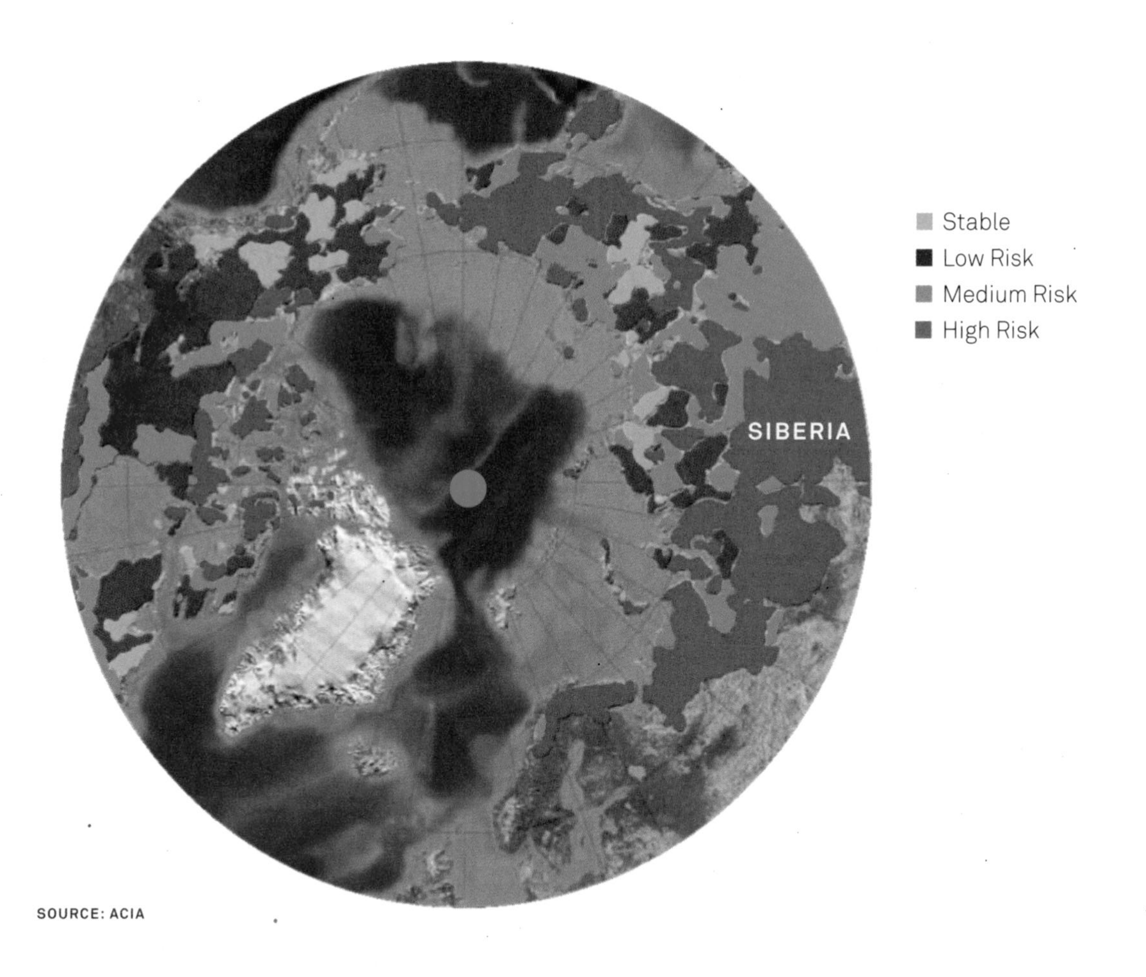

chapter seven

The Ends of the Earth: The South Pole

ANTARCTICA IS THE closest thing to another planet we can experience on this one. It is surreal—completely and unremittingly white in every direction. It is vast and cold—much colder than the Arctic. And in contrast to the thin layer of ice that is the Arctic, the ice at the Antarctic ice cap is 10,000 feet (almost two miles) thick.

The enormity of all that snow and ice masks a surprising fact: Antarctica is actually a desert. It meets the technical definition because it receives less than one inch of precipitation per year. Think about it—an icy desert.

Significant numbers of penguins, seals, and birds hug the edge of Antarctica and find food in the ocean. But a little farther inland, there are absolutely no signs of life—other than small groups of scientists who usually do not venture too far or too long from their heated research labs.

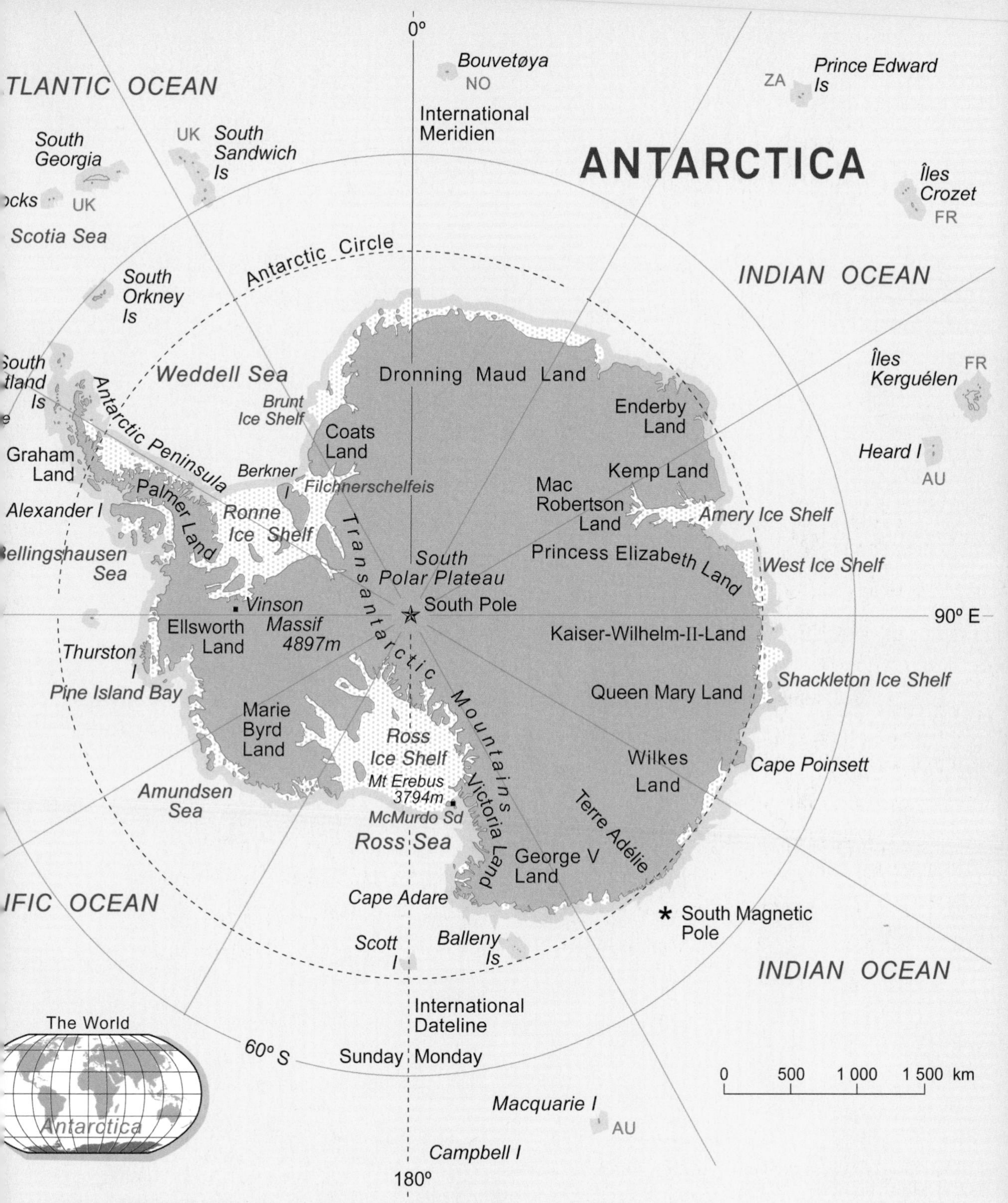

Source: Composite Gazetteer of Antarctica, © Her Majesty the Queen in Right of Canada, Natural Resources of Canada, 2001

The documentary film *March of the Penguins* was a surprise hit in 2005. However, the movie neglected to point out that the population of emperor penguins is thinning.

Since the 1970s, the penguins' neighborhood has become increasingly warm. The Southern Ocean experiences natural shifts in weather from one decade to the next, but this warm spell has continued, causing the thinning of sea ice. Less sea ice means fewer krill, the penguins' main food source. Also, the weakened ice is more likely to break apart and drift out to sea, carrying off the young penguin chicks, who often drown.

Is global warming responsible for the thinning penguin population? Scientists believe so.

Emperor penguin family,
Weddell Sea, Antarctica

In 1978 a scientist warned that the sign of a dangerous warming trend in the Antarctic would be the breakup of its ice shelves.

The enlarged map is of the Antarctic Peninsula jutting off the west coast of the continent. Each orange splotch represents an ice shelf the size of Rhode Island or larger that has broken up in the years since the scientist's warning.

This is the Larsen-B ice shelf. On the map on the facing page, it is the maroon area on the peninsula. The ice rises roughly 700 feet above the ocean surface—that's as high as a seventy-story skyscraper.

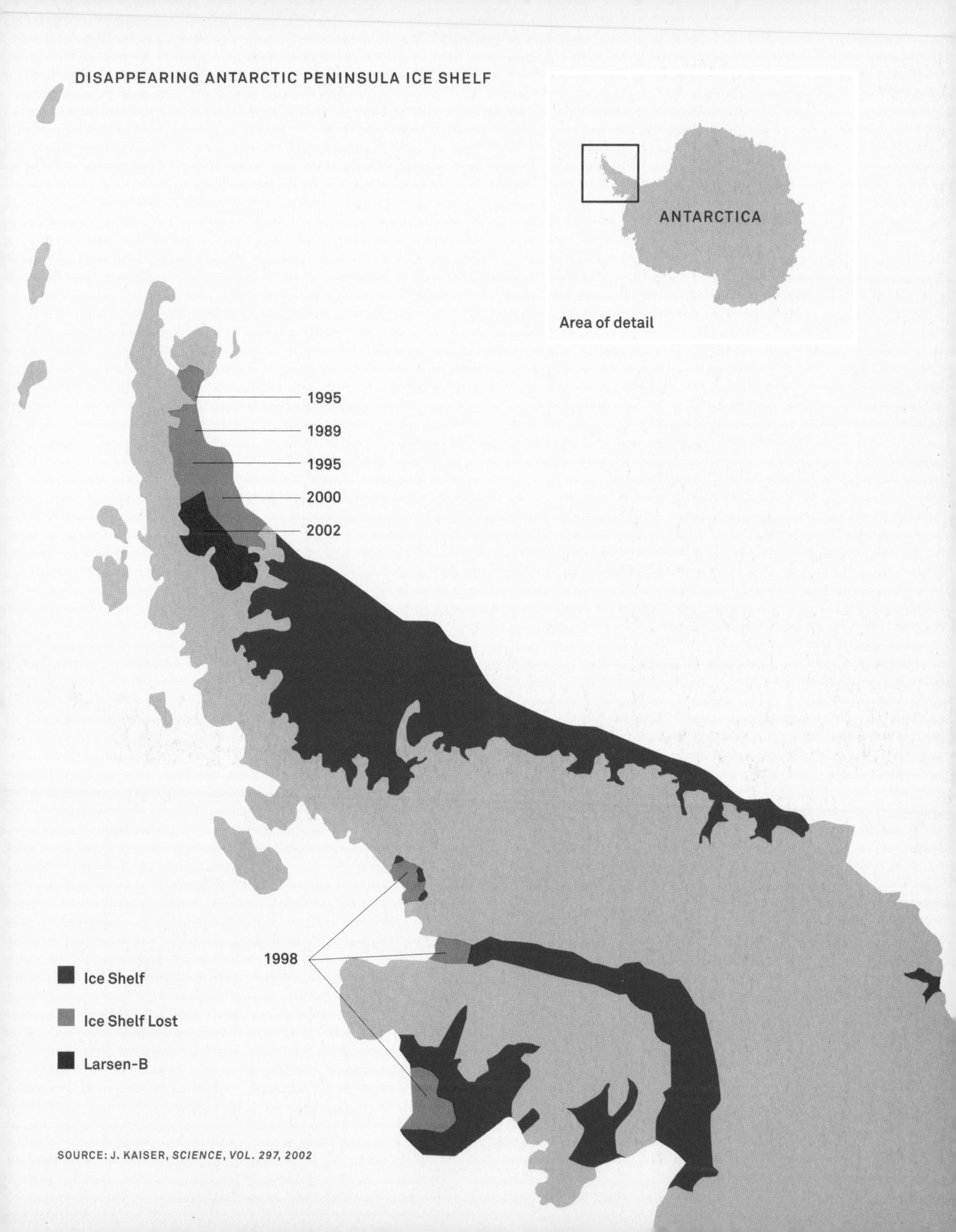
DISAPPEARING ANTARCTIC PENINSULA ICE SHELF
ANTARCTICA
Area of detail
1995
1989
1995
2000
2002
1998
Ice Shelf
Ice Shelf Lost
Larsen-B
SOURCE: J. KAISER, *SCIENCE*, *VOL. 297*, *2002*

SOURCE: J. KAISER. *SCIENCE*, 2002

Satellite image of the Larsen-B ice shelf, January 31, 2002

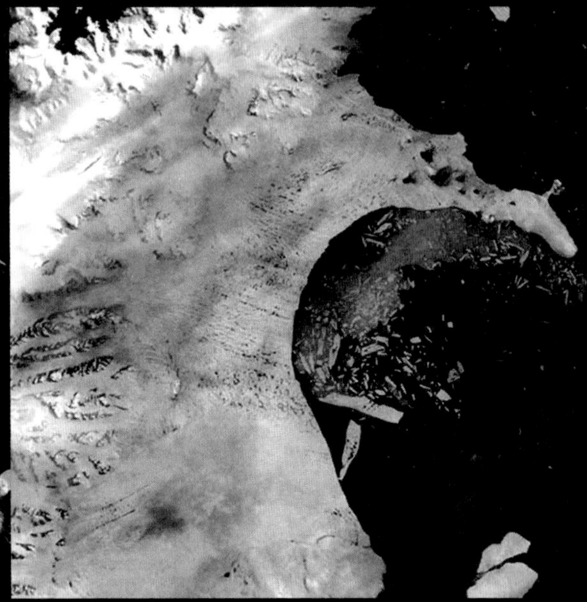

February 17, 2002

The Larsen-B ice shelf was about 150 miles long and 30 miles wide in 2002. Scientists thought it would be stable for at least another century—even with global warming. But starting on January 31, 2002, it completely broke up within 35 days. Scientists were shocked.

What they hadn't taken into account was the effect of melting pools of water on top of the ice mass. (You can see them on the images as black streaks against the white.) Scientists had mistakenly thought this meltwater sank back into the ice and refroze. But it wasn't doing that. It kept sinking straight down, boring holes, making the ice shelf look like Swiss cheese, weakening the entire mass.

SOURCE: MODIS IMAGES COURTESY OF NASA'S TERRA SATELLITE

February 23, 2002

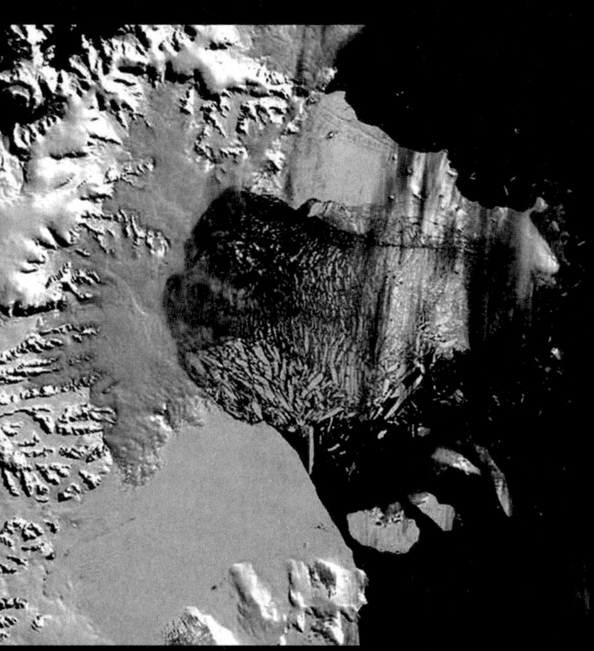

March 5, 2002

Once a sea-based ice shelf breaks apart, the land-based ice held back behind it becomes unstable. Then, when the ice that was on land falls into the sea, it raises the sea level. This is one of the reasons sea levels will continue to rise worldwide unless global warming is checked.

Ice breaking off the front of a glacier, 1995

In East Antarctica, there is an ice sheet that is the largest ice mass on the Earth. It has always seemed to be increasing in size, but now there are scientific studies that seem to show it is shrinking, and most of the glaciers there now appear to be moving faster toward the sea.

The two 2006 studies that showed this also measured air temperatures high above the ice. Temperatures there have gone up faster than air temperatures anywhere else in the world. This surprised the scientists, and they aren't sure exactly how to explain it.

EAST ANTARCTICA
WEST ANTARCTIC
ICE SHELF
ANTARCTICA

Although geographically most of Greenland lies within the Arctic Circle, physically it resembles Antarctica. Just about the same size as the present West Antarctic ice shelf, Greenland is land topped with a dome of ice that on average is 5,000 feet thick. The West Antarctic ice shelf is propped up on the tops of islands, which means that if it melted or slipped into the sea, it would raise sea levels worldwide by twenty feet. Sea levels would also rise by twenty feet if Greenland melted or broke up and slid into the sea.

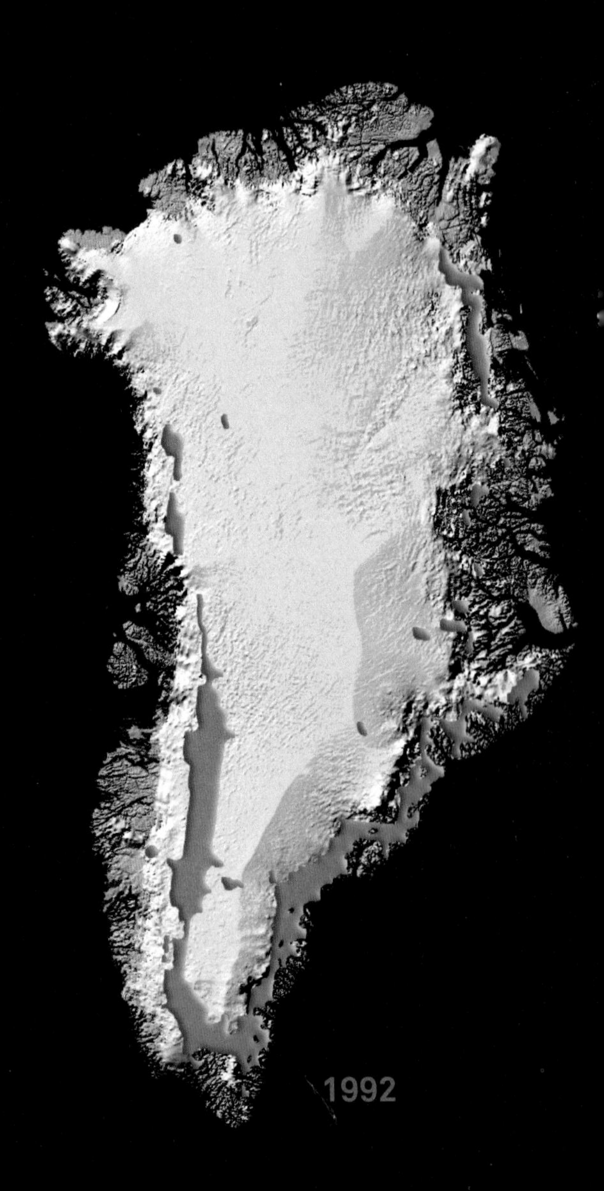

The red areas on these maps show the amount of melting in Greenland in 1992, then ten years later in 2002, and finally only three years after that, in 2005.

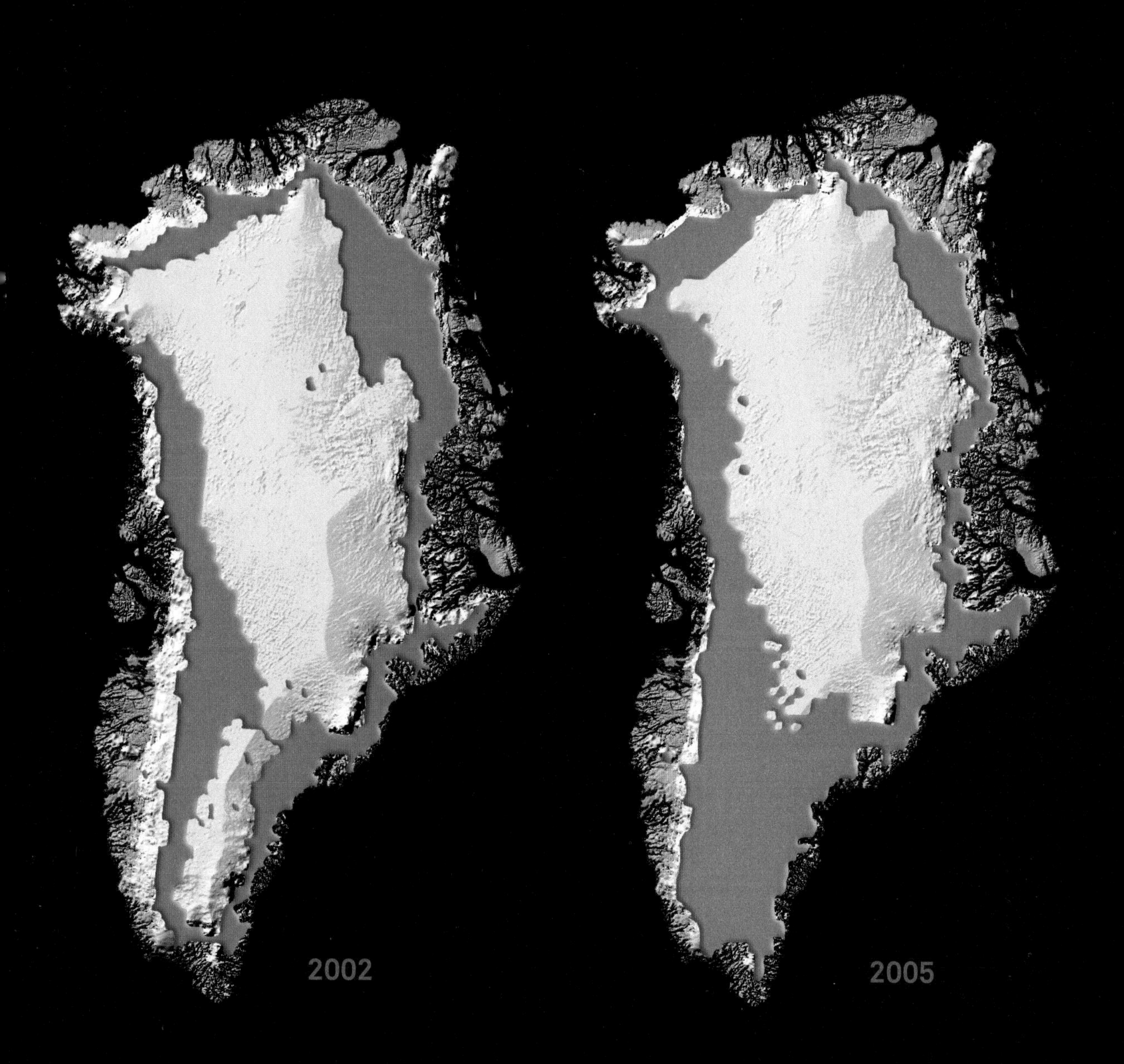

SOURCE: ©2005 ACIA

I have visited Greenland twice. In 2005 I flew over Greenland and saw for myself the pools of meltwater. These pools have always been there, but the difference now is that there are so many more of them in more places. This is important because these are the same kinds of meltwater pools that scientists saw on top of the Larsen-B ice shelf before it suddenly disappeared. In Greenland, just as in the Antarctic Peninsula, scientists now know the meltwater is sinking to the bottom and cutting deep crevasses and tunnels, called "moulins," which can make the ice mass much less stable and might cause it to slide more quickly toward the ocean. There has always been some seasonal melting in Greenland, and moulins have formed in the past, but in recent years, the melting has been happening faster, and all year round.

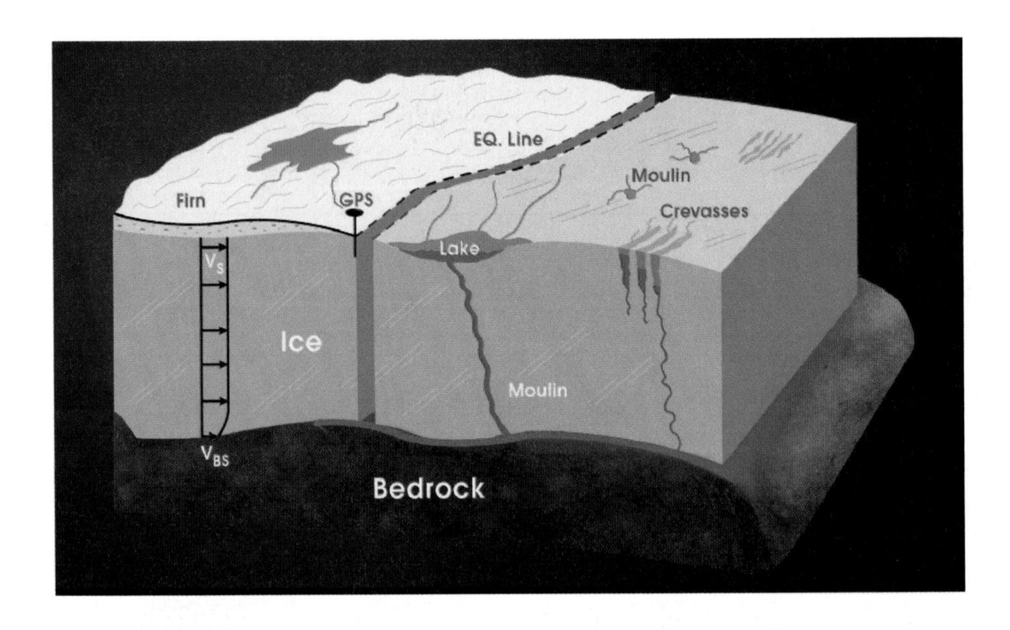

Stream of meltwater cascading off
the ice sheet, Greenland, 2005

chapter eight

A New Atlas?

High tide in Funafuti, Tuvalu, Polynesia

IF GREENLAND'S ICE DOME or the West Antarctic ice shelf melted or broke up into the sea, it would raise water levels worldwide between 18 and 20 feet.

Rising seas would mean millions of people would have to evacuate their homes. This has already happened in low-lying island nations in the Pacific.

It would also mean that maps of the world would have to be redrawn.

This is what would happen to Florida. Miami would be underwater.

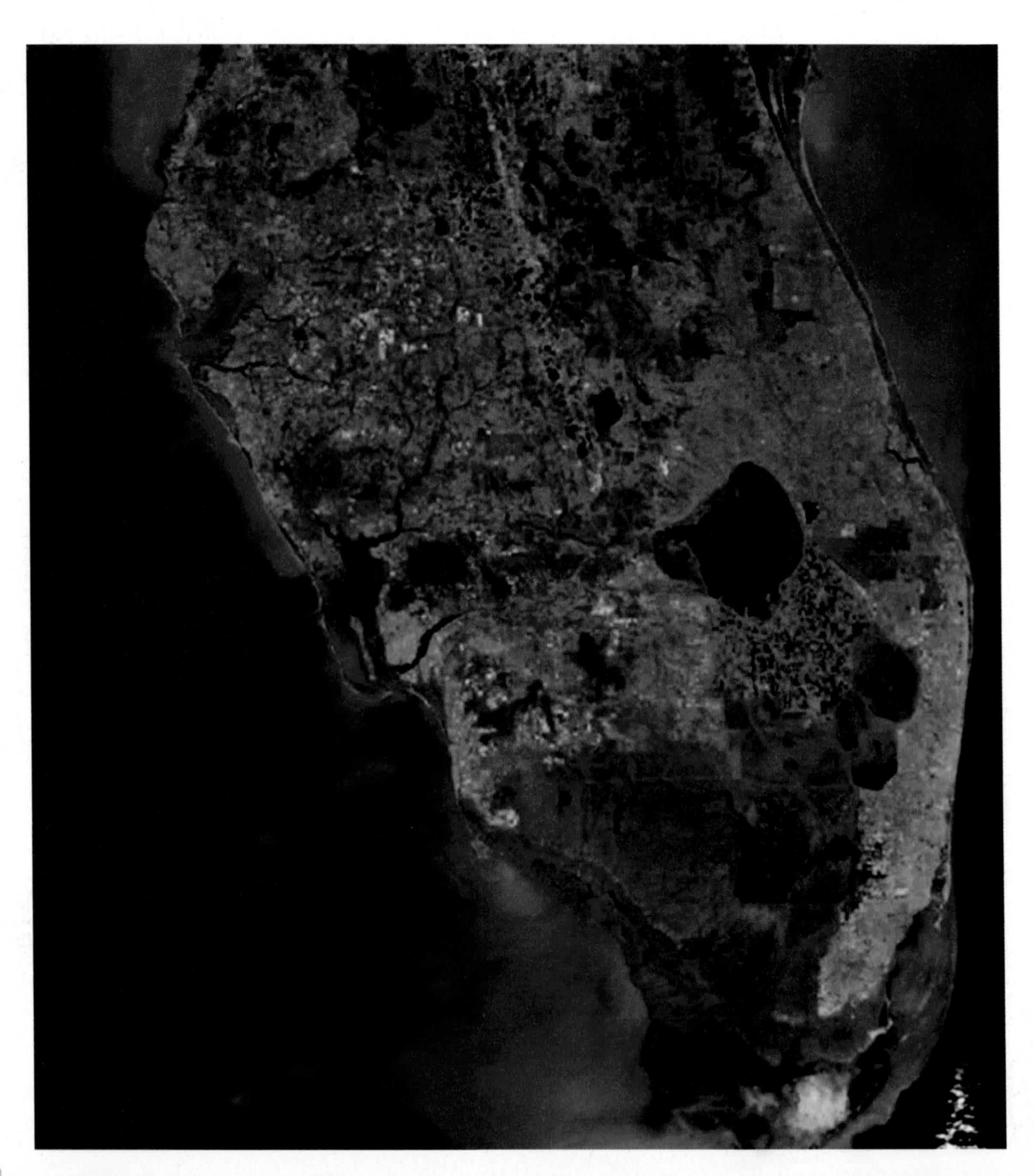

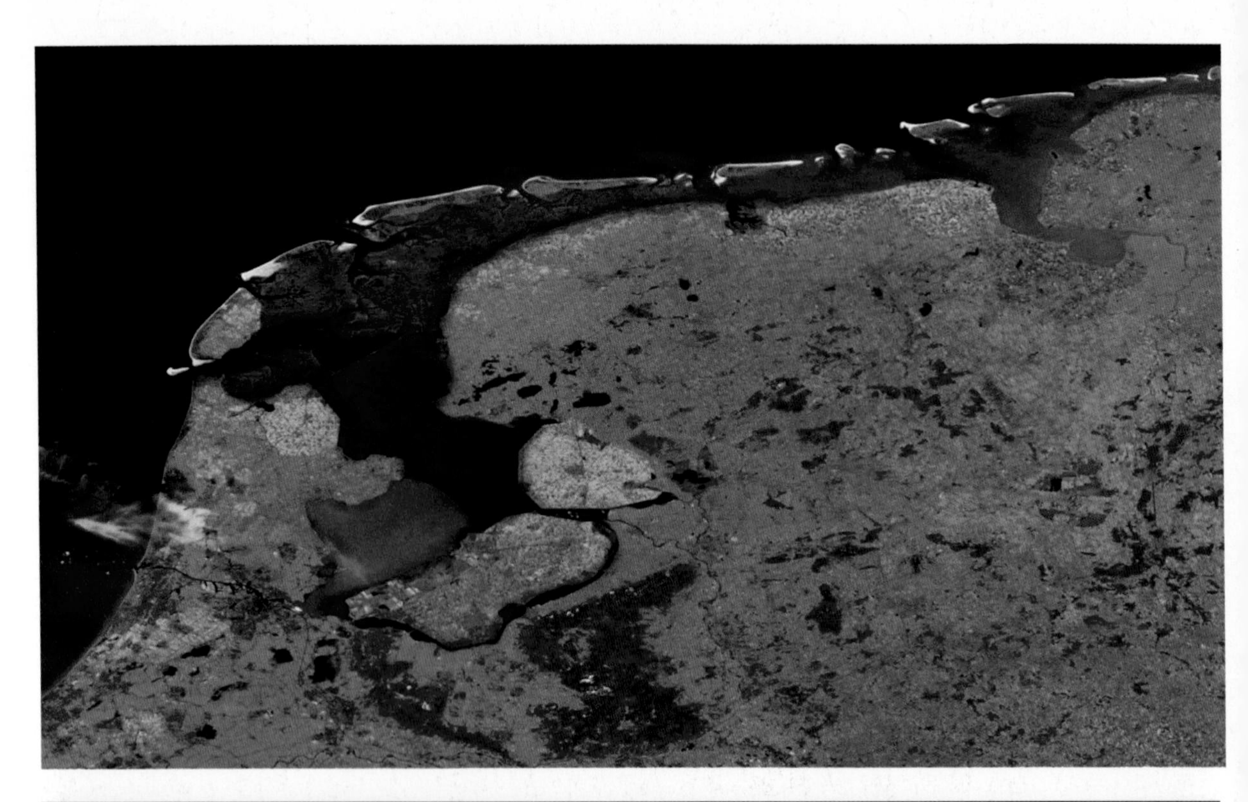

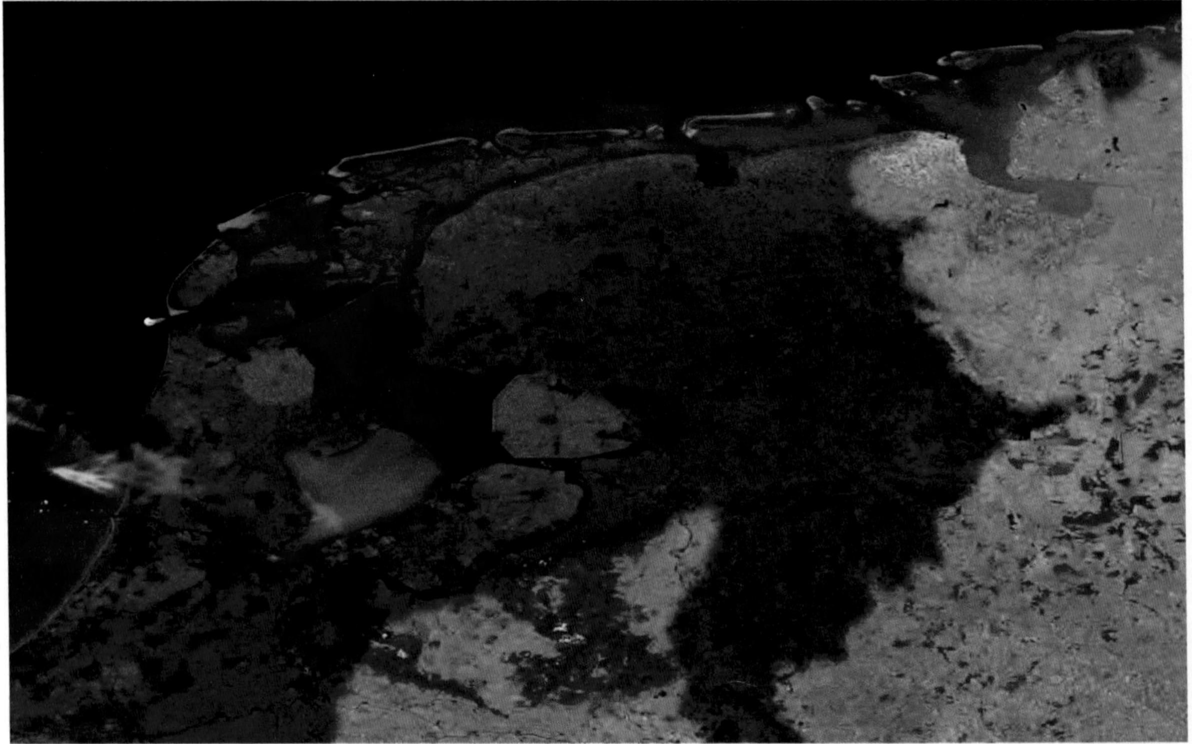

This is what would happen to the Netherlands, a low-lying nation that faces the North Sea.

Amsterdam would vanish. The Dutch have always had to find ways to contend with the problem of flooding, because almost a third of the country is actually below sea level.

A few years ago, they held a competition among architects to design floating homes.

Floating houses, Amsterdam, the Netherlands, 2000

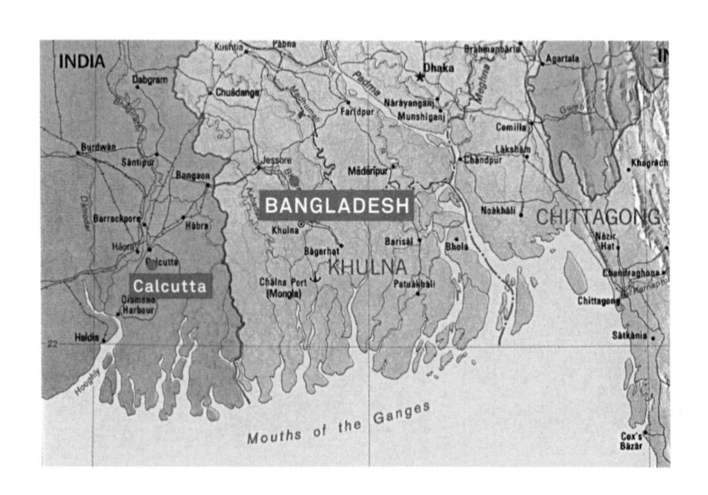
INDIA
BANGLADESH
Calcutta
KHULNA
CHITTAGONG
Mouths of the Ganges
Dhaka
Jessore
Khulna
Barisal
Chittagong

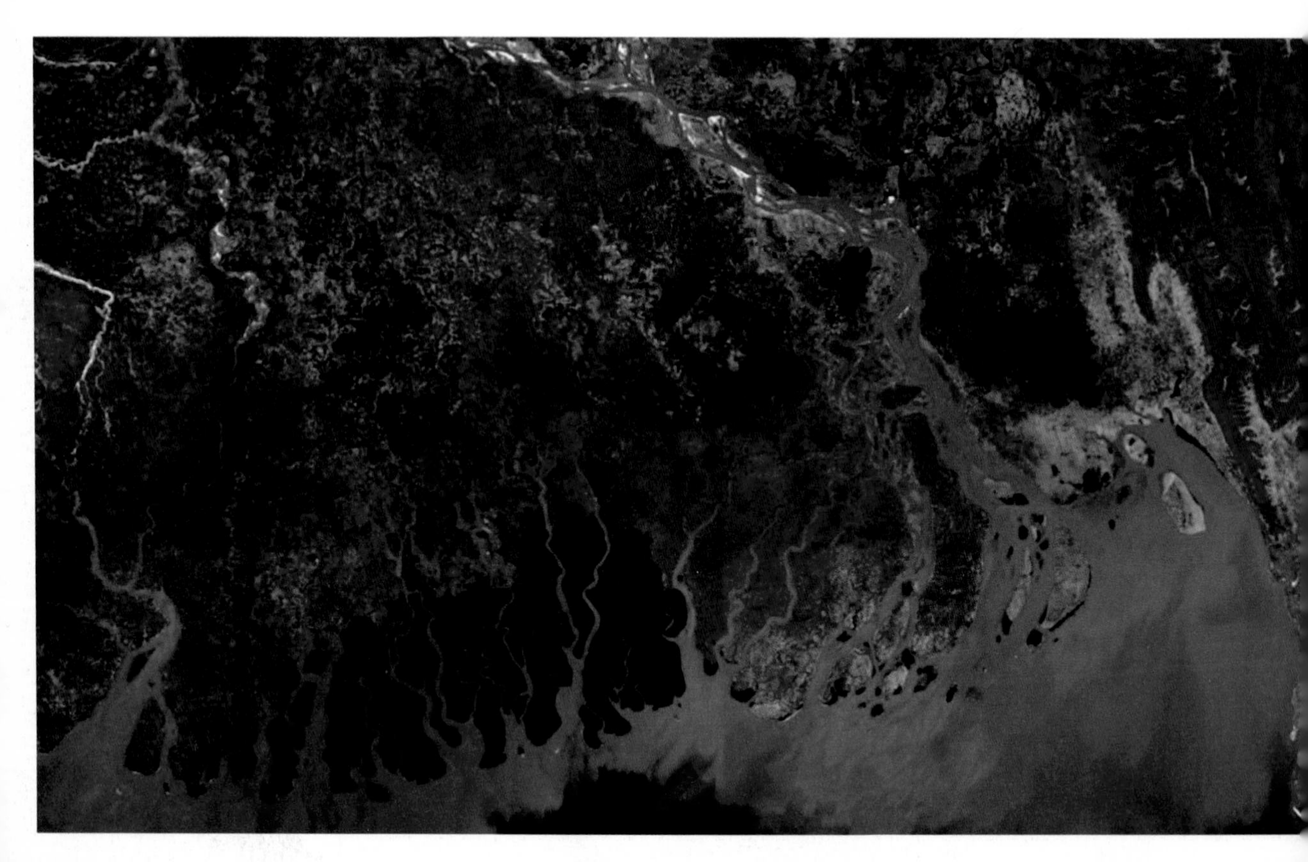

In Bangladesh and the city of Calcutta, 60 million people would be uprooted. That's roughly the entire population of France or the U.K. or Italy.

Manhattan is an island. At its southern end, where the World Trade Center once stood, new buildings will go up, as well as a museum and memorial to the people who lost their lives there on September 11, 2001.

If sea levels rose twenty feet worldwide, the site of the World Trade Center Memorial would be underwater.

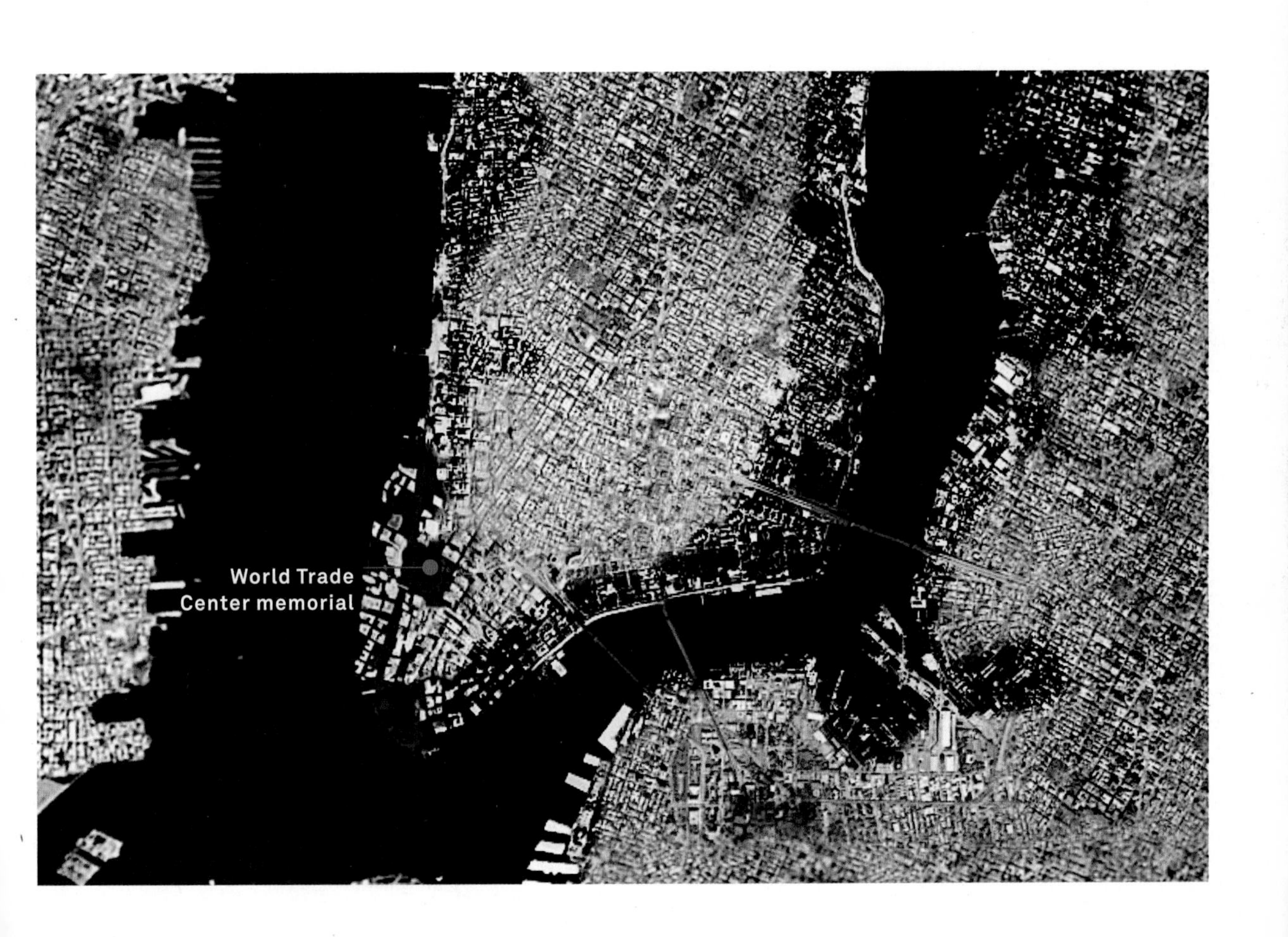

chapter nine

Deep Trouble

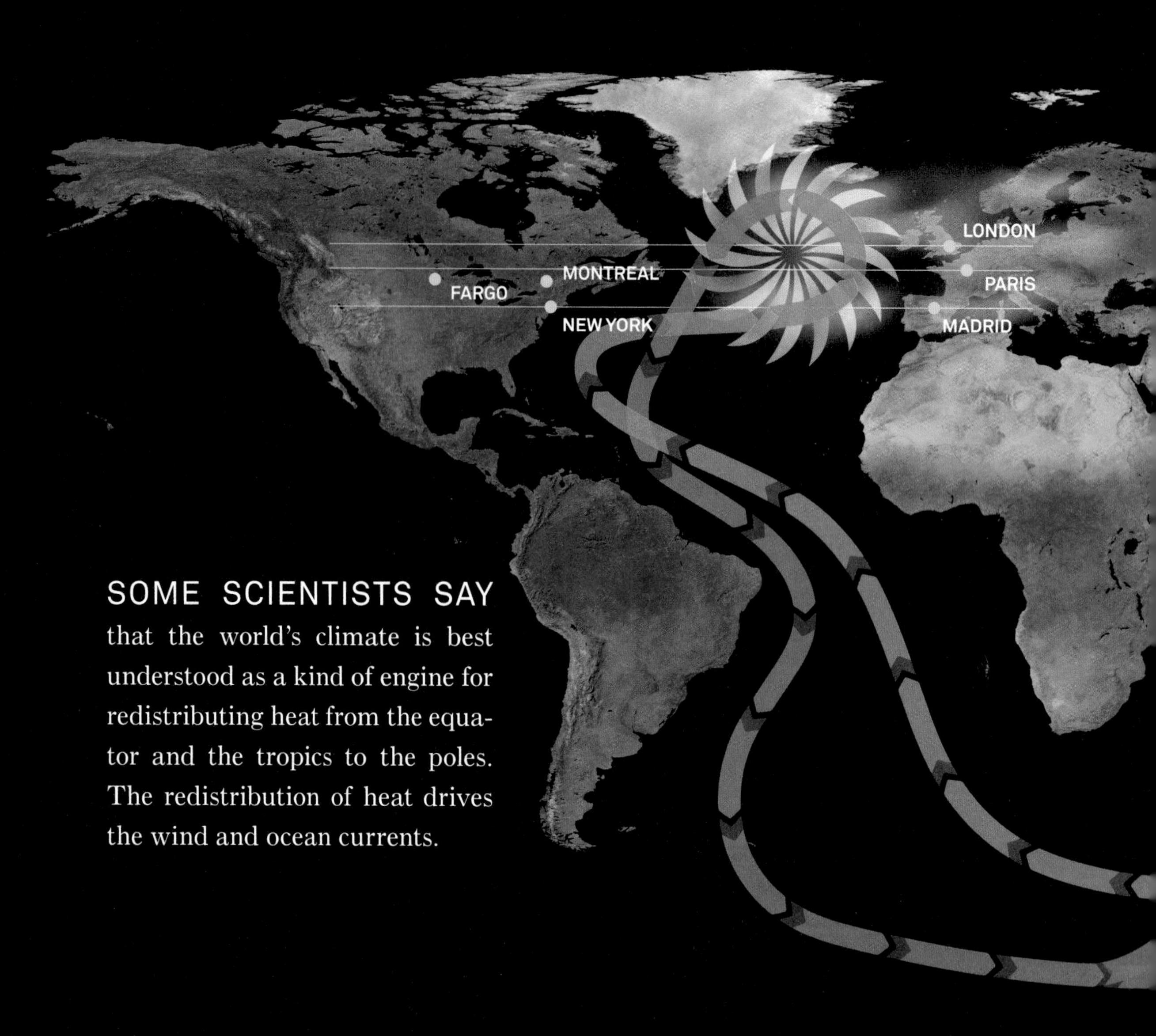

SOME SCIENTISTS SAY that the world's climate is best understood as a kind of engine for redistributing heat from the equator and the tropics to the poles. The redistribution of heat drives the wind and ocean currents.

The different ocean currents of warm and cold water are all linked in a big loop called the Global Ocean Conveyor Belt, shown here. The red parts of the loop represent warm surface currents, the best known of which is the Gulf Stream. The blue parts represent deep cold-water currents flowing in the opposite direction.

Some scientists think global warming could potentially disrupt the workings of this conveyor belt, with disastrous consequences to the climate worldwide. If the conveyor belt stops moving, some areas would be too cold and others too hot.

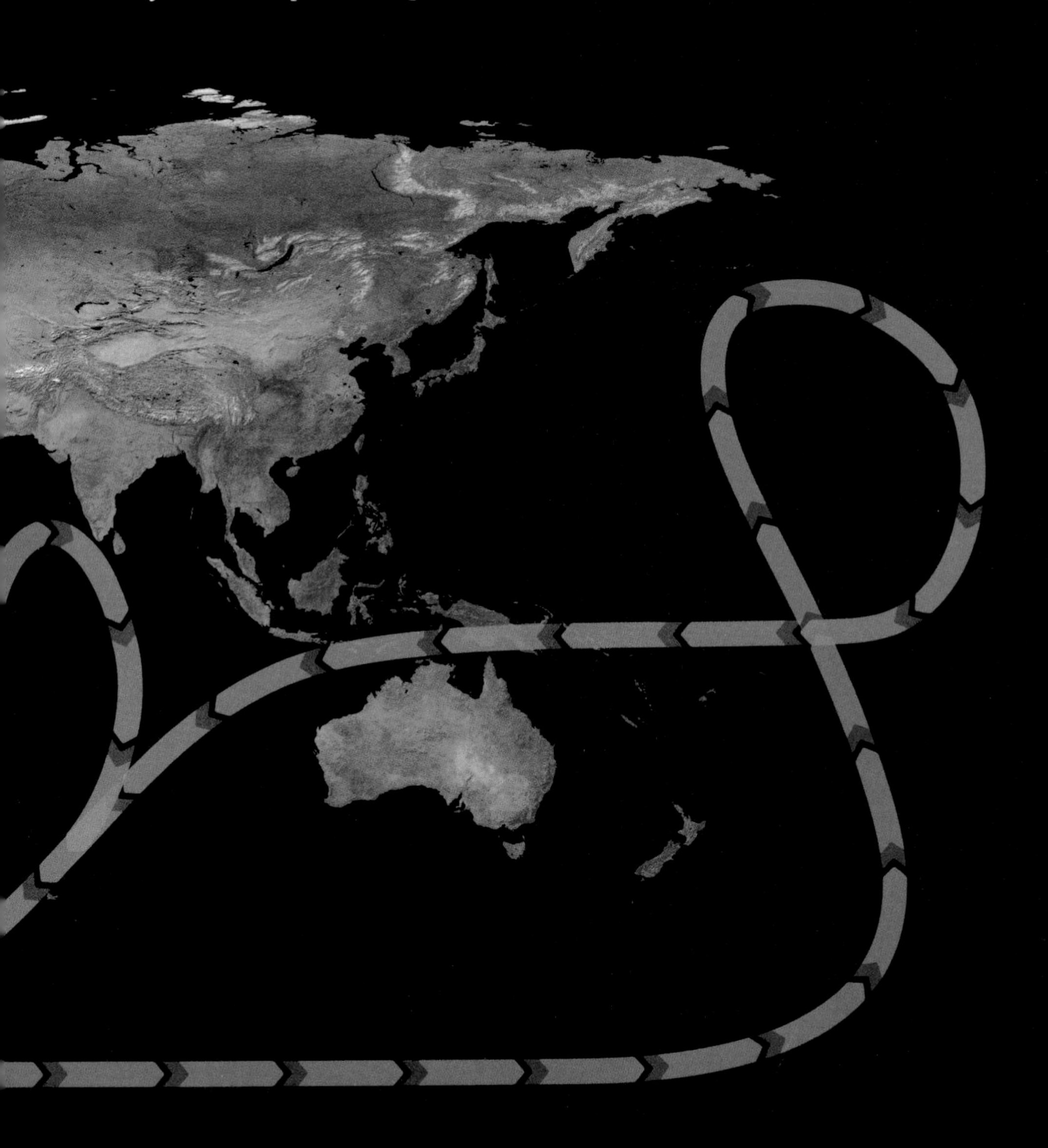

Coral reefs, which are as important to ocean species as rainforests are to land species, are being killed in large numbers by global warming.

Many factors contribute to the death of coral reefs—pollution from nearby shores, destructive dynamite fishing, and more acidic ocean waters. However, the most deadly cause of the recent, rapid, and unprecedented deterioration of coral reefs is believed by scientists to be higher ocean temperatures due to global warming.

Lettuce coral, Phoenix Islands, Kiribati, Polynesia, 2004

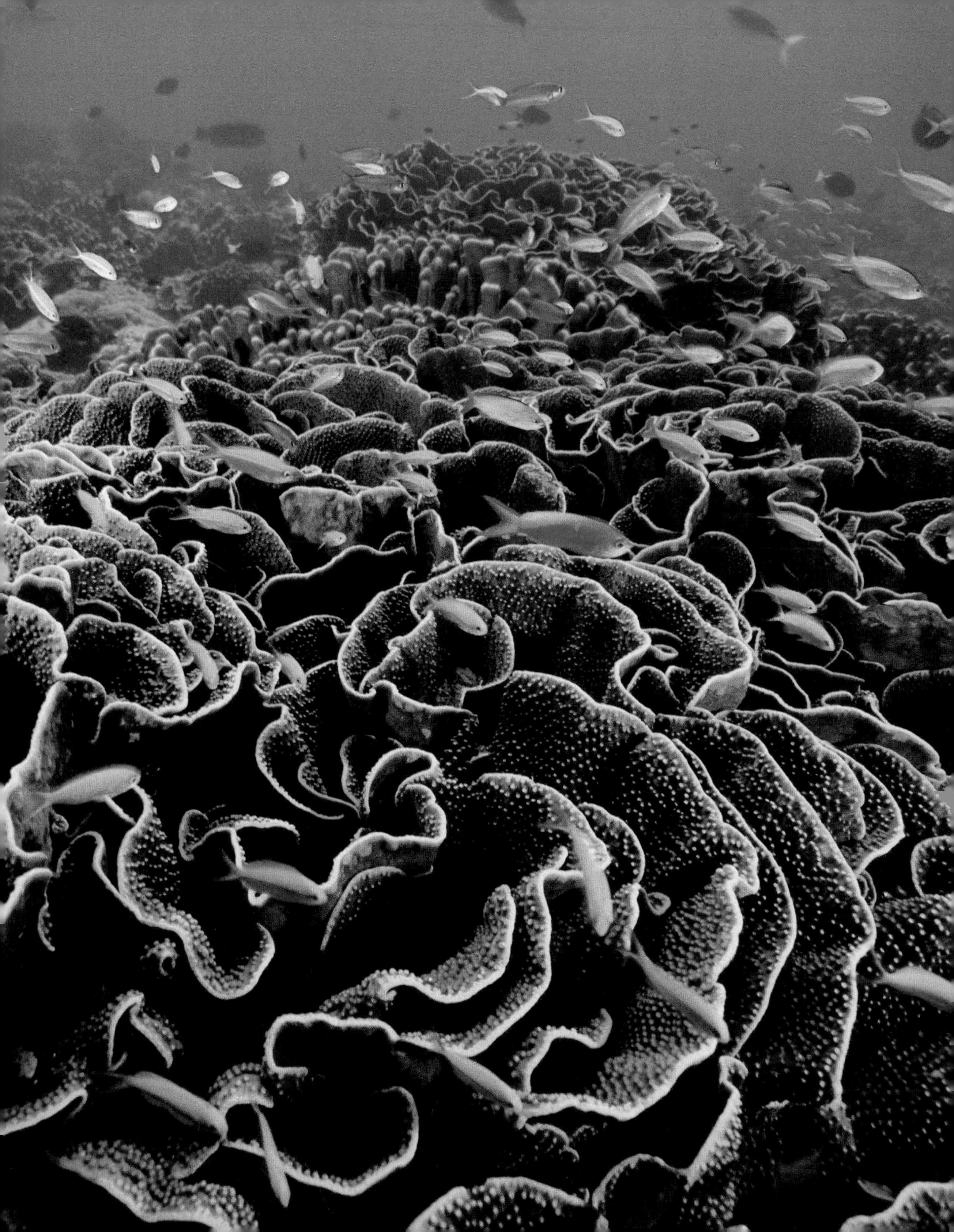

Why?

"Zooks" is the nickname for the algae called zooxanthellae that live in coral. Zooks not only account for the coral's beautiful color, but they also provide nutrients for it. But when the ocean water gets too warm, the coral expels them. Once they are gone, the coral is no longer colorful or healthy, and all you see is the ghostly white or gray skeleton of the reef beneath. This transformation, known as bleaching, is a warning signal that the coral reef is about to die.

In 1998, the second hottest year on record, the world lost an estimated 16 percent of its coral reefs.

Ten to fifteen years ago, the link between global warming and the large-scale bleaching of corals was considered controversial. Now it is universally accepted.

Bleached coral, Rongelap Reef, Marshall Islands, 2004

chapter ten

Hazardous to Your Health

WE ARE CHANGING the chemistry of our oceans in many ways. This leads to new "dead zones," where the water has too little oxygen to support most forms of life. Some dead zones are caused by the appearance of algae blooms.

An algae bloom is an explosion of algae that occurs in warmer waters and is fed by pollution from human activity on the shore. Sometimes they are poisonous—Florida's red tide is one example. Algae blooms occur naturally, but in recent years they have grown to spectacular sizes never seen before. In the Baltic Sea in northern Europe, many resorts had to close during the summer of 2005 because of the algae.

Algae bloom in the Baltic Sea, Gotland, Sweden, 2005

As the climate gets warmer, disease-carrying species move into new places.

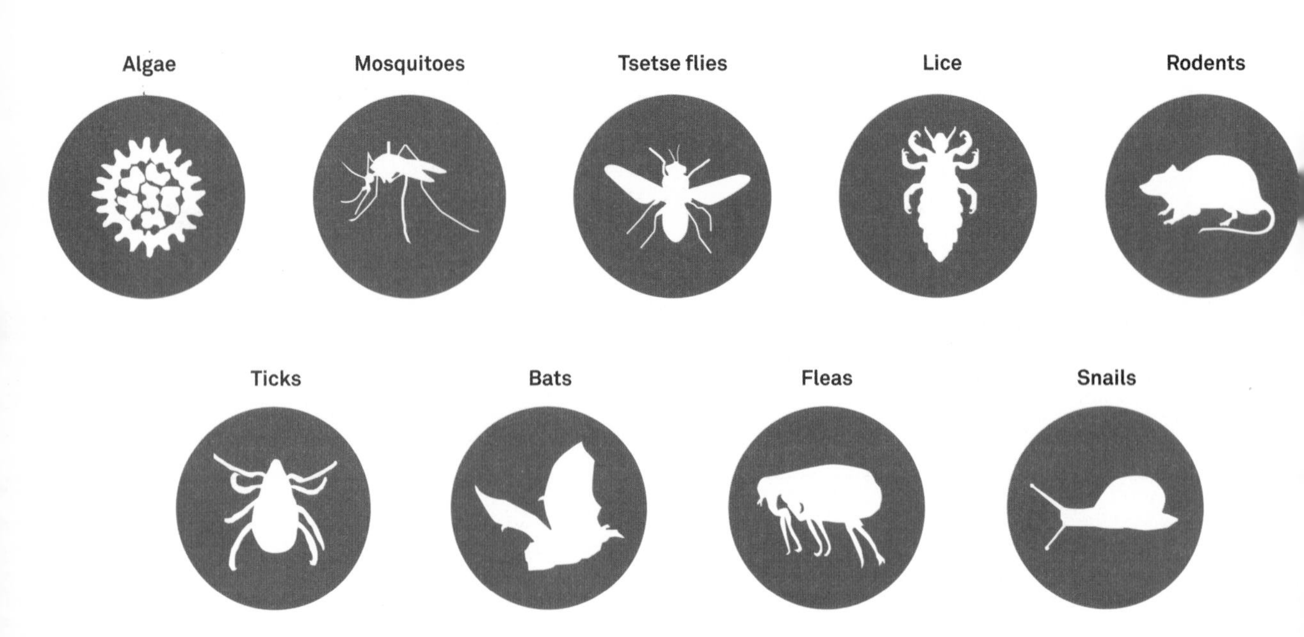

In general, the world of germs and viruses is less threatening to human beings when there are colder nights and colder winters, and when the climate of an area remains stable. The cold weather keeps germs at bay. Global warming, on the other hand, increases our vulnerability to diseases.

Mosquito

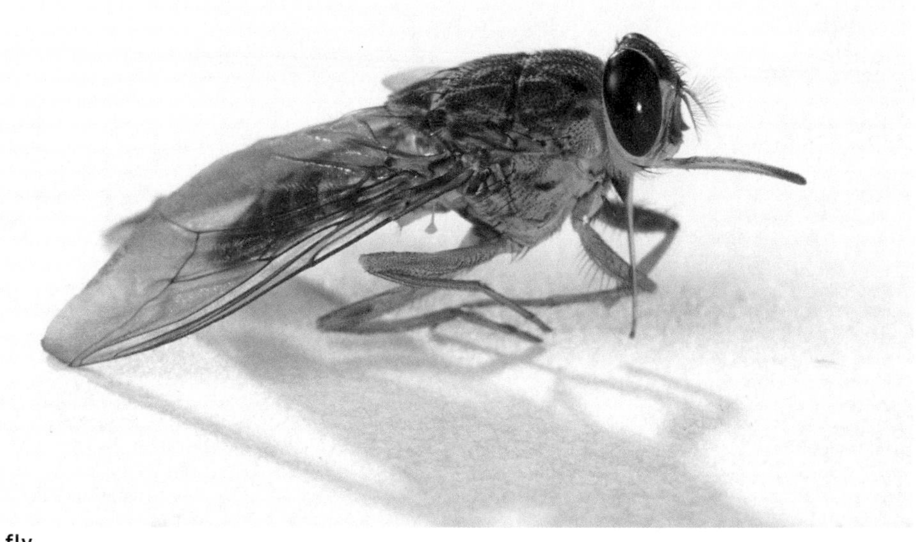

Tsetse fly

Mosquitoes spread many diseases, such as malaria, encephalitis, West Nile virus, and yellow fever. They breed and live in warm climates. Because of global warming, they are now traveling to places where they've never been before.

For instance, in Africa, the city of Nairobi, the capital of Kenya, and the city of Harare, the capital of Zimbabwe, used to be above the mosquito line (the highest point at which mosquitoes can live).

This picture illustrates how mosquitoes have moved to higher elevations.

MOSQUITOES MOVE TO HIGHER ELEVATIONS

West Nile virus had never entered the United States until 1999. Then within two years it crossed the Mississippi River. Two years after that, West Nile spread all the way across the continent. It is not usually a serious illness, but in rare cases, it can be fatal. Scientists speculate that the unusual heat and rain levels during the years 1999 to 2003 contributed to the swift spread of the mosquitoes that transmit the virus.

SPREAD OF WEST NILE VIRUS IN THE UNITED STATES

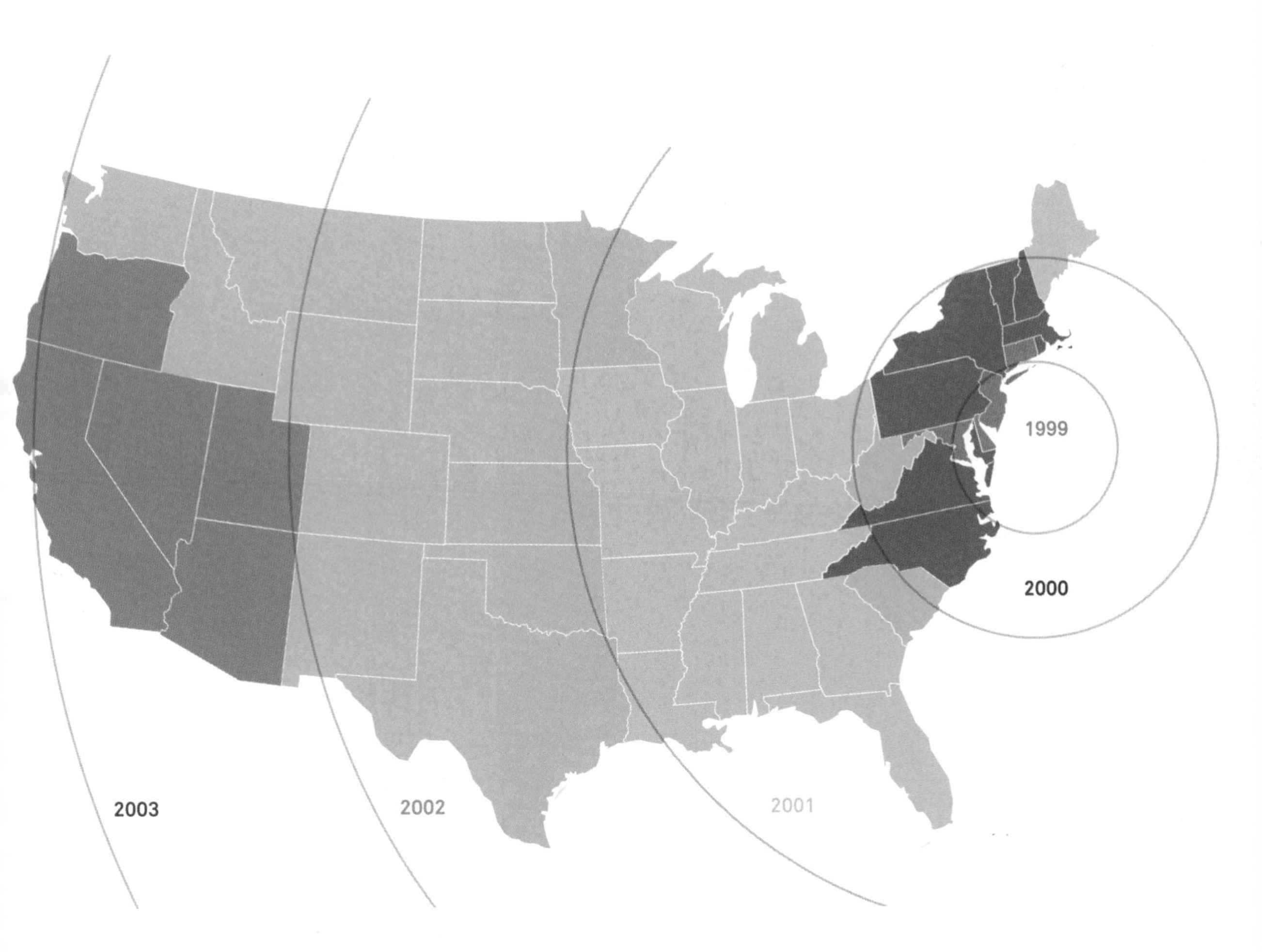

SOURCE: COMPILED FROM CDC, HEALTH CANADA, USGS, AND PROMED-MAIL SOURCES AS OF MAY 14, 2003

chapter eleven

Off Balance

GLOBAL WARMING IS throwing off the age-old rhythm of the Earth's seasons. In some places, the length of the days and the temperature are getting out of synch. That is, the days may be short, as they always are in winter, but the temperature is unusually high and springlike. Because of climate imbalances like this, many species around the world face new challenges. The pied fly-catcher is one example. A study from the Netherlands explains what has happened.

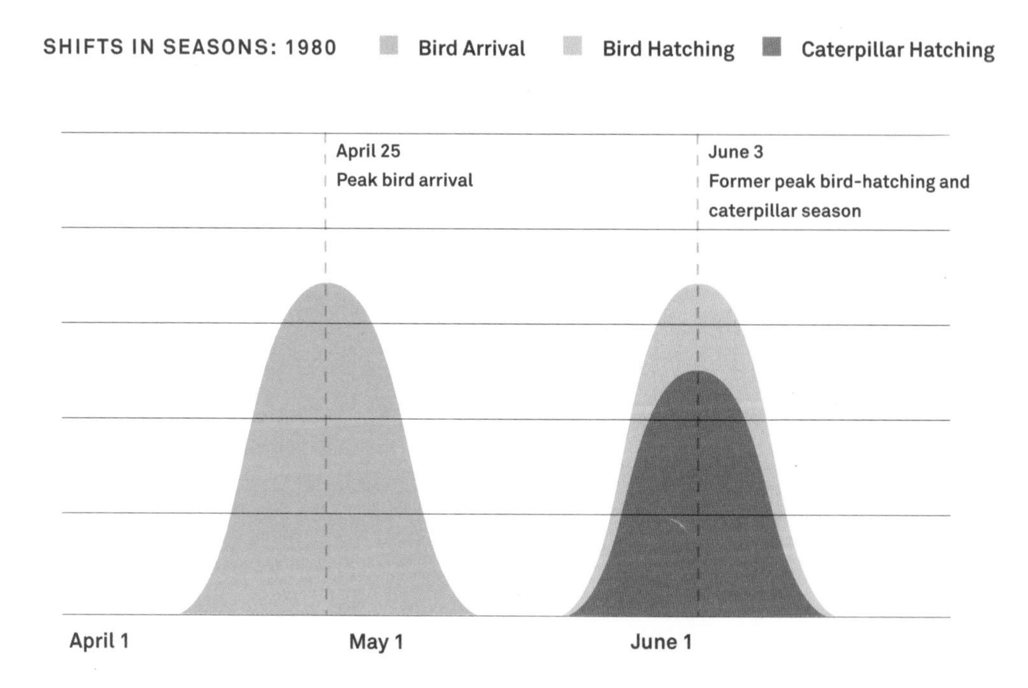

Pied fly-catchers are migratory birds. Twenty-five years ago, the peak arrival date for pied fly-catchers in the Netherlands was April twenty-fifth. Chicks hatched almost six weeks later, peaking on June third. This was also the height of caterpillar season. Perfect timing, as caterpillars are the traditional source of food for the chicks.

Mother bird feeding its young, the Netherlands

But now the schedule has been thrown off. The birds still arrive in late April. However, after more than two decades of global warming, the caterpillars are peaking two weeks earlier. May fifteenth is their new peak date. The chicks' peak hatching date has also moved up a little—to May twenty-fifth—but not early enough to coincide with the greatest number of caterpillars. As a result, it's harder for the mother birds to find food for their young.

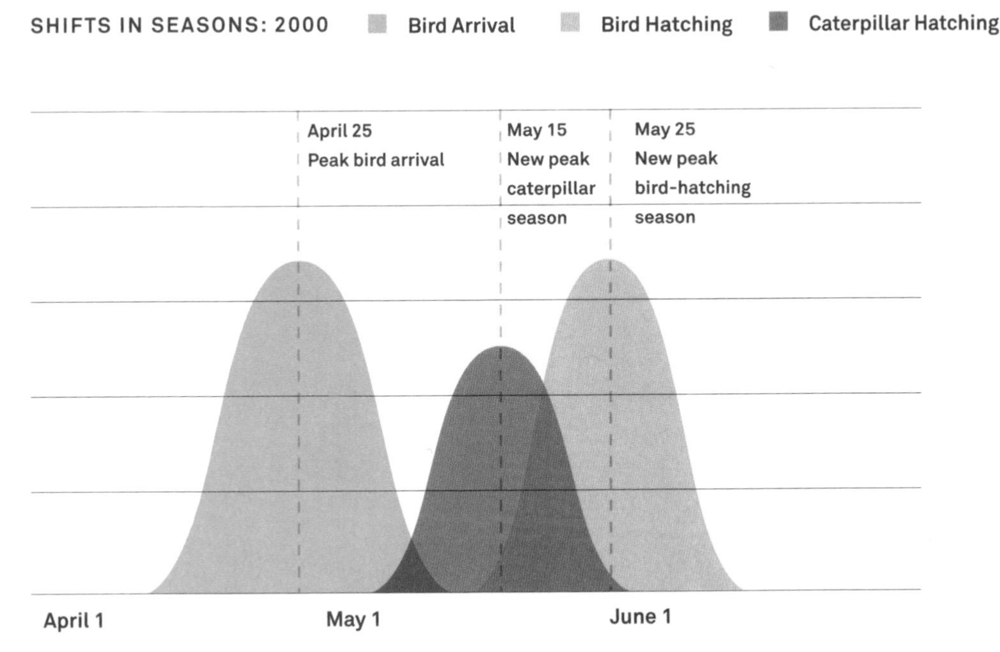

SOURCE: *SCIENTIFIC AMERICAN*

Here is another example of how global warming disrupts the balance of nature as we have known it.

In the American West, colder winters used to slow the destructive spread of pine beetles. These insects bore into tree bark to lay eggs, eventually killing off the tree. Now there are fewer days of frost, and the pine beetles are out in force. The photograph shows the damage they do.

Pine beetle damage, Plains, Montana, 1989

Giant Glass Frog

Greater Mouse Lemur

White-Fronted Goose

Bowhead Whale

Grey-Headed Albatross

Emperor Penguin

Golden Toad

Macaroni Penguin

Coqui (tree frog)

Flightless Cormorant

Antarctic Fur Seal

Wattled Cranes

Yellow-Eyed Penguin

Polar Bear

Red-Breasted Goose

Leopard Seal

Many species around the world are now threatened with extinction, in part because of the climate crisis and in part because of what humans are doing to places where these creatures once thrived.

Actually the two reasons are interwoven: for example, cutting down trees in the Amazon rain forest (something humans are responsible for) destroys habitats and drives species to extinction, while at the same time the felled trees release more CO_2 into the atmosphere, causing further threats to species' survival.

It's a spiraling cycle of destruction.

chapter twelve

Collision Course

WE ARE WITNESSING an unprecedented and massive collision between our civilization and the Earth. We are trashing the planet.

How has this happened?

One major reason is that there are so many more of us on the Earth.

Refuse dump in Mexico City, Mexico, 1996

World population has skyrocketed.

When the baby boom generation that I'm part of was born at the end of World War II, the population had already crossed the two billion mark; in my lifetime, I've watched it go all the way to six and a half billion. And by 2050, there will be over nine billion people.

In many ways this is a success story. Death rates are going down everywhere in the world. Birth rates are also decreasing—especially in the most advanced countries. And because the average family size is shrinking, the rate of population increase is slowing. But we have quadrupled world population in less than a hundred years. Our impact on the Earth is greater now.

We have a moral obligation to take into account this dramatic change in terms of how we treat the planet.

POPULATION GROWTH THROUGHOUT HISTORY

First modern humans

160,000 BC 100,000 BC 10,000 BC 7000 BC 6000 BC 5000 BC 4000 BC

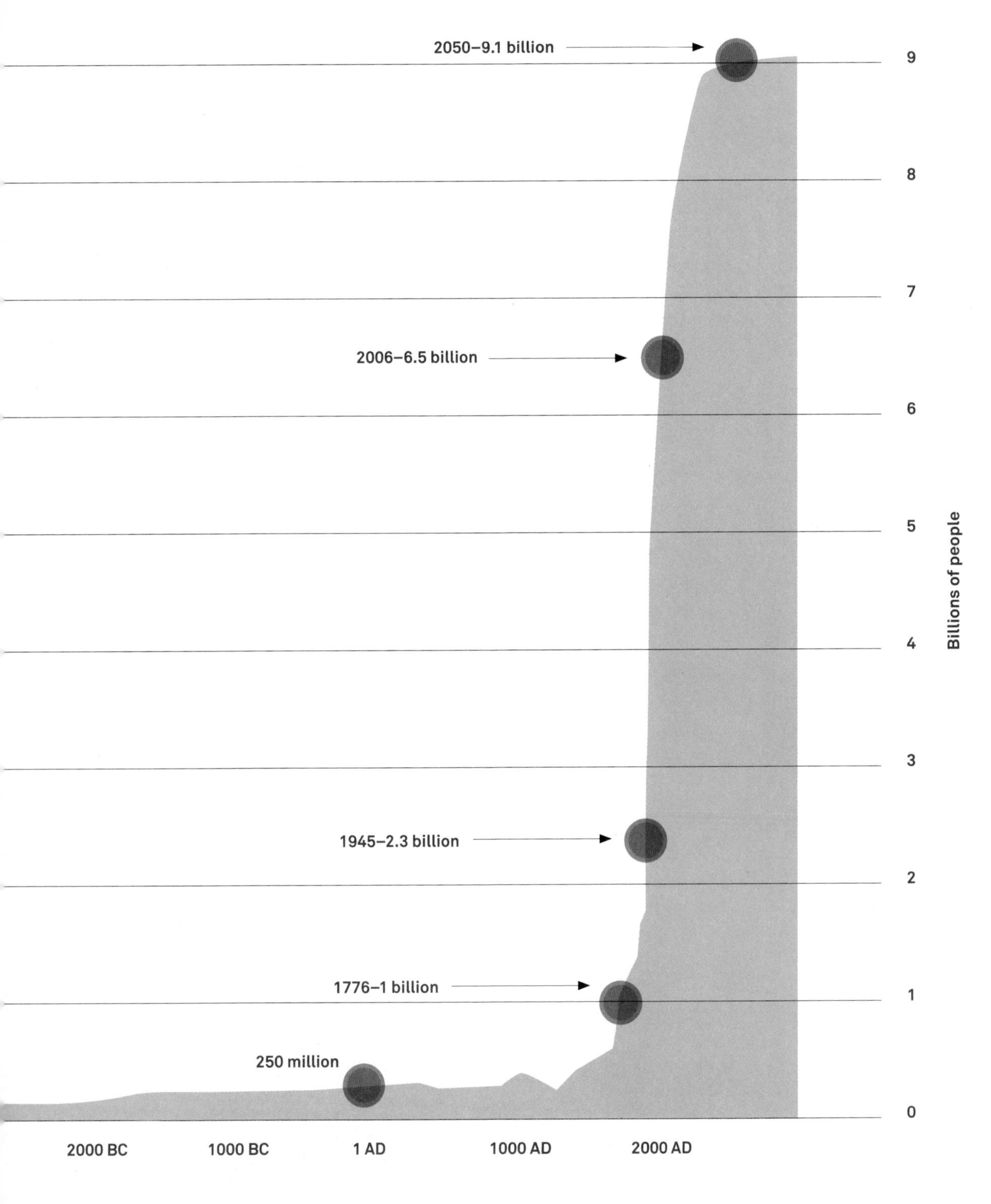

SOURCE: UNITED NATIONS

Around the world, cities have become home to enormous numbers of people.

Shinjuku district of Tokyo, Japan, 1996

The population of the Tokyo metropolitan area, the largest in the world, has grown to more than 35 million.

Logging, Tapajos National Forest, Brazil, 2004

Because of this rapid rise in population, there's greater demand worldwide for food, shelter, water, and energy, which in turn puts a strain on all our natural resources.

And although the greatest proportion of people lives in cities, vast areas of remote forestland are cut down to provide timber and grow food for city dwellers.

The photo on the opposite page shows a forest in Washington State after clearcutting, which means removing all the trees and undergrowth.

Stumps and slash after clearcutting near Forks, Washington, 1999

Perhaps no place on Earth is in greater jeopardy than the Amazon tropical rain forest. It covers an immense area—over 2.5 million square miles. But a large part of it is being cut down and destroyed. Some trees are cut down for logging or large-scale ranching. Much of the forest destruction comes from burning. Almost 30 percent of the CO_2 released into the atmosphere each year is a result of wood fires used for cooking, and the burning of brushland for subsistence farming (whereby poor families cut down a few acres to grow enough food to survive).

Clearing the rain forest has an immediate effect—cutting down trees releases carbon dioxide. It also has a significant long-term effect—the trees that have always been there to suck up carbon dioxide are gone for good.

What happens locally has worldwide consequences.

Farm worker clearing land for ranching, Rondonia, Brazil, 1988

HAITI

The countries of Haiti and the Dominican Republic occupy the same island in the Caribbean. Haiti's government, however, has done nothing to safeguard the environment, and 98 percent of their forests have been cut down.

The Dominican Republic has a different policy. They protect their forests. Look at the contrast between the two countries.

DOMINICAN REPUBLIC

Composite satellite view of the Earth at night, 1994–1995

chapter thirteen
Technology's Side Effects

TECHNOLOGY HAS IMPROVED people's lives in many ways. A hundred and fifty years ago, before the invention of the electric lightbulb, the world at night was dark.

SOURCE: NASA

A Vanishing Act

But technological power hasn't always been used wisely. For example, irrigation has long worked wonders for humankind. But now we have the ability to divert giant rivers according to our own design, instead of nature's.

There is danger when too much water is diverted from its natural source.

The former Soviet Union diverted water from two mighty rivers in central Asia in order to irrigate cotton fields. These rivers had fed the Aral Sea.

When I went to the site of the Aral Sea some years ago, I saw a strange sight: an enormous fishing fleet marooned in the sand, with no water in sight. The entire Aral Sea was gone. This picture shows part of that grounded fleet.

Grounded fishing boats, Aral Sea, Kazakhstan, 1990

The way we now access copper through strip-mining ravages the land.

Copper strip mine,
Cannea, Mexico, 1993

New technologies for fighting, such as nuclear bombs, have dramatically changed the consequences of war.

German soldiers in World War I, 1914

Test detonation of a nuclear bomb, Nevada, 1957

The most technologically advanced countries have the greatest obligation to use technology wisely and treat the planet responsibly. Yet the country best-known for technological might is the one most responsible for the problem of global warming.

The United States.

Yes, our country emits more greenhouse gas pollution than South America, Africa, the Middle East, Australia, and Asia all put together.

On these two pages, countries and continents are pictured in relative size, according to what percentage of greenhouse gases they emit. For example, Australia, which is not much smaller than the United States in terms of land, appears tiny because it only accounts for 1.1 percent of the polluting gases released.

We must seek changes.

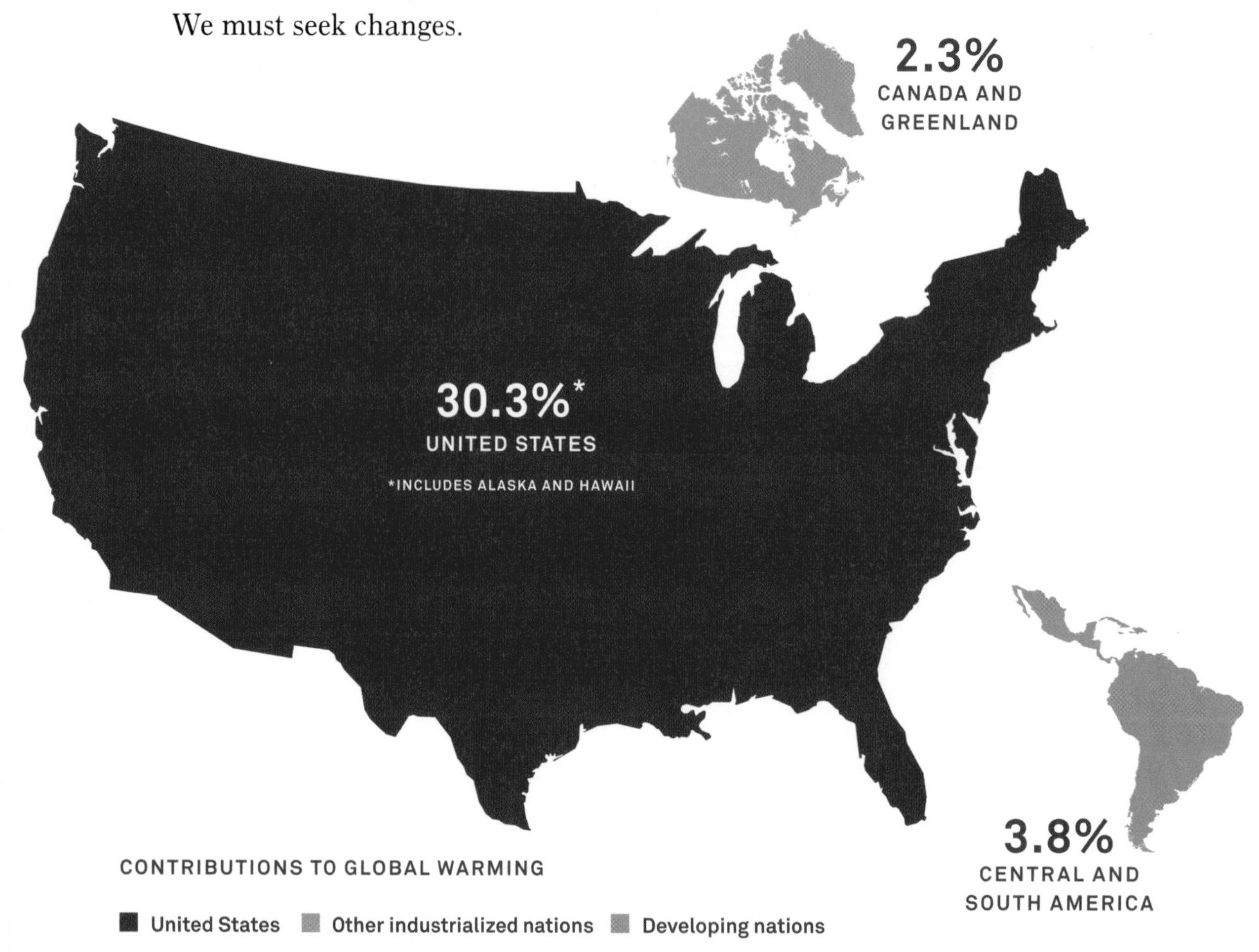

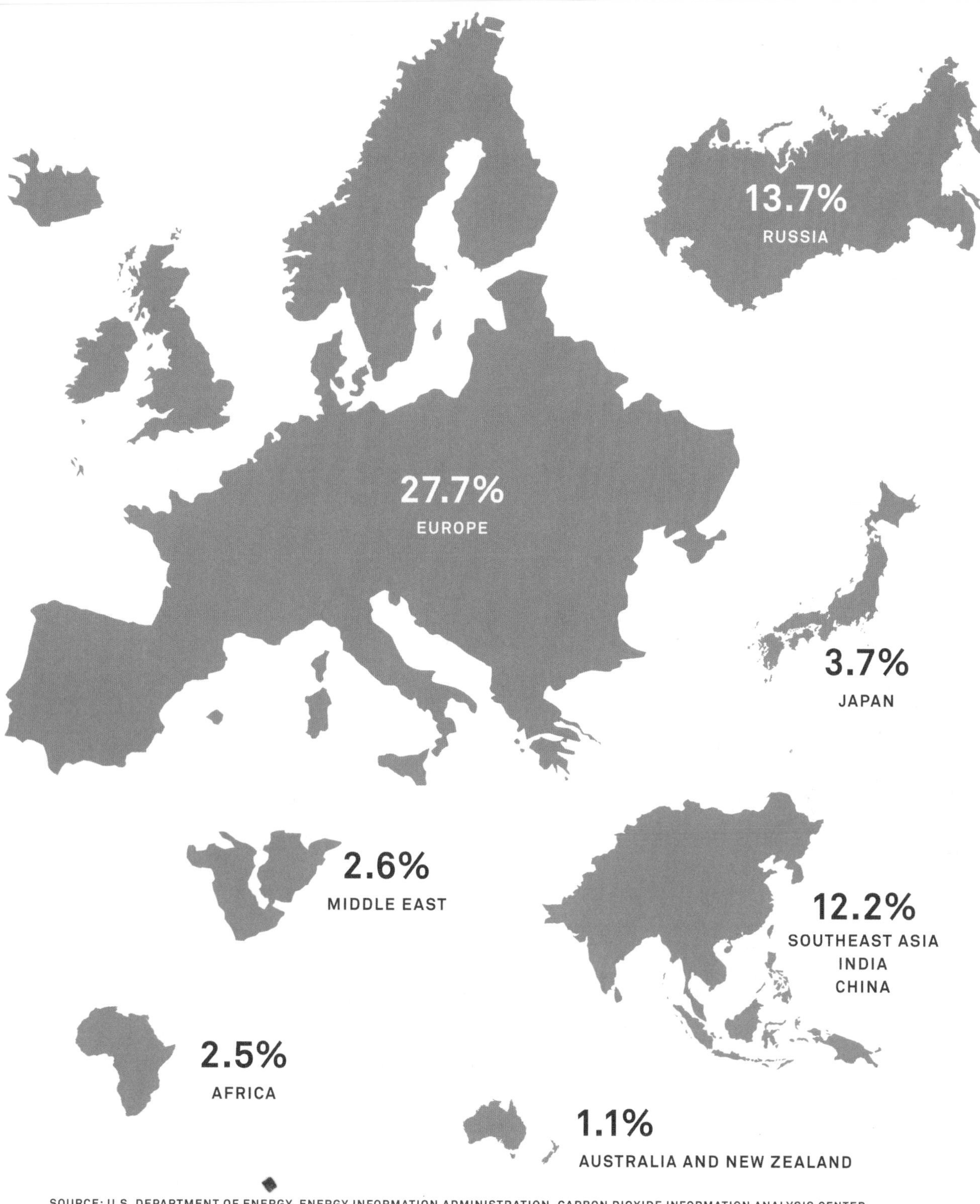

SOURCE: U.S. DEPARTMENT OF ENERGY, ENERGY INFORMATION ADMINISTRATION, CARBON DIOXIDE INFORMATION ANALYSIS CENTER

chapter fourteen

"Denial Ain't Just a River in Egypt"

MARK TWAIN'S WORDS above sum up another reason why we remain on a collision course with nature. Some people don't want to admit there is a climate crisis. Others trust false and misleading information from political groups that want to confuse people into thinking there is no problem.

I can think of another case when this happened: cigarettes. In the 1960s, scientific research proved smoking was gravely harmful to our health, the cause of heart disease and cancer. The tobacco industry's response was to conduct a "disinformation" campaign to confuse people and make them doubt the facts.

Cigarette packages are now required by law to display the Surgeon General's Warning

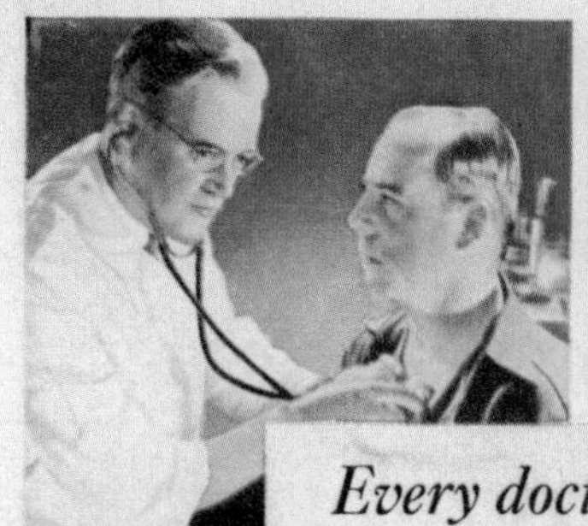

Every doctor in private practice was asked:
—family physicians, surgeons, specialists . . .
doctors in every branch of medicine—
"What cigarette do you smoke?"

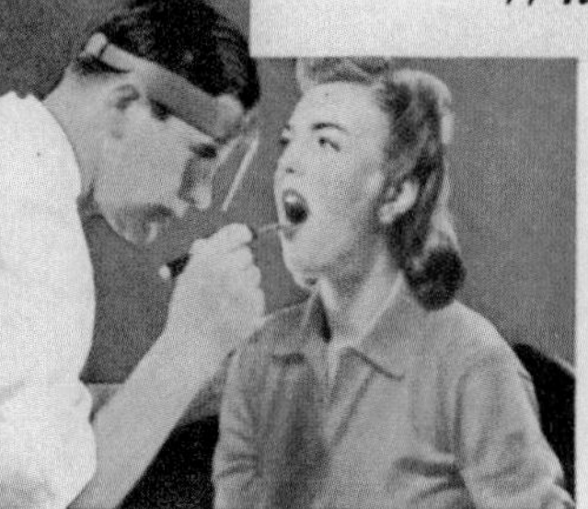

According to a recent Nationwide survey:

More Doctors Smoke Camels

than any other cigarette!

Not a guess, not just a trend . . . but an *actual fact* based on the statements of doctors themselves to 3 nationally known independent research organizations.

THE "T-ZONE" TEST WILL TELL YOU

The "T-Zone"—T for taste and T for throat—is your own laboratory, your proving ground, for any cigarette. For only your taste and your throat can decide which cigarette tastes best to *you* . . . and how it affects your throat. On the basis of the experience of many, many millions of smokers, we believe Camels will suit your "T-Zone" to a "T."

CAMEL
CHOICE QUALITY
TURKISH & DOMESTIC BLEND CIGARETTES

R. J. Reynolds Tobacco Co.
Winston-Salem, N. C.

YES, your doctor was asked . . . along with thousands and thousands of other doctors from Maine to California.

And they've named their choice—the brand that more doctors named as their smoke is *Camel!* Three nationally known independent research organizations found this to be a fact.

Nothing unusual about it. Doctors smoke for pleasure just like the rest of us. They appreciate, just as you, a mildness that's cool and easy on the throat. They too enjoy the full, rich flavor of expertly blended costlier tobaccos. Next time, try Camels.

An advertisement from the early 1960s promoting smoking

Scientists are in almost complete agreement about the causes of global warming.

The University of California reviewed all of the expert scientific studies of global warming for the last ten years, and when they looked at a large random sample, they didn't find a single one that disagreed with the mainstream view.

Number of peer-reviewed articles dealing with "climate change" published in scientific journals during the previous 10 years:

928

Percentage of articles in doubt as to the cause of global warming:

0%

Yet certain companies that don't want to control their global warming pollution pressure our government to let them conduct business as usual. These self-interest groups mount campaigns to throw evidence into question.

In articles about global warming, newspapers often report both "sides" of the story and give each side (scientists and self-interest groups) equal weight. This gives the false impression that there *is* a debate over facts when there isn't. No wonder people are confused.

Articles in the popular press about global warming during the previous 14 years:

3,543

Percentage of articles in doubt as to the cause of global warming:

53%

There are powerful businesses that make money from activities that worsen global warming. They want to censor scientific research and mask the truth about the crisis we are facing. In cases where these industries have connections to politicians, "skeptics" have been appointed to important government environmental posts. These "skeptics" prevent action to reduce global warming. It's like appointing a fox to guard the henhouse.

In 2001, a man named Philip Cooney was appointed to a position giving him responsibility for the environmental policy of the White House.

What was his background for the job?

For the previous six years, Mr. Cooney had been in charge of the oil industry's campaign to confuse the public about global warming.

Here you see part of a 2005 memo in which Mr. Cooney censored information supplied by the Environmental Protection Agency. He took out all mention of the dangers that global warming poses to the American people. But before the memo—an official White House memo—was released in the censored form, it was leaked to *The New York Times*, which disclosed what Mr. Cooney was doing. He resigned and the next day he went to work for Exxon Mobil.

The marked-up White House memo below was leaked to The New York Times

~~Warming will also cause reductions in mountain glaciers and advance the timing of the melt of mountain snow peaks in polar regions. In turn, runoff rates will change and flood potential will be altered in ways that are currently not well understood. There will be significant shifts in the seasonality of runoff that will have serious impacts on native populations that rely on fishing and hunting for their livelihood. These changes will be further complicated by shifts in precipitation regimes and a possible intensification and increased frequency of hydrologic events.~~ Reducing the uncertainties in current understanding of the relationships between climate change and Arctic hydrology is critical.

straying from
research
strategy in
speculative
findings fro
here.

Philip Cooney

1995–JANUARY 20, 2001

American Petroleum Institute lobbyist in charge of global warming disinformation

JANUARY 20, 2001

Hired as chief of staff, White House Environment Office

JUNE 14, 2005

Leaves White House to go on payroll of Exxon Mobil

Other countries are taking action to reduce harmful gases released into the atmosphere. One way is by setting higher auto mileage standards so less gas will be burned. In Japan (the dark blue line on the graph), cars are required by law to get more than 45 miles per gallon. By 2012 the European Union (the green line) will surpass Japanese standards, requiring 52 miles per gallon. China requires cars to get more than 35 miles per gallon. Our friends in Canada (the light blue line) and Australia (the yellow line) are moving toward higher requirements.

California (the orange line) has set its own standards, but overall the United States (the red line) is dead last.

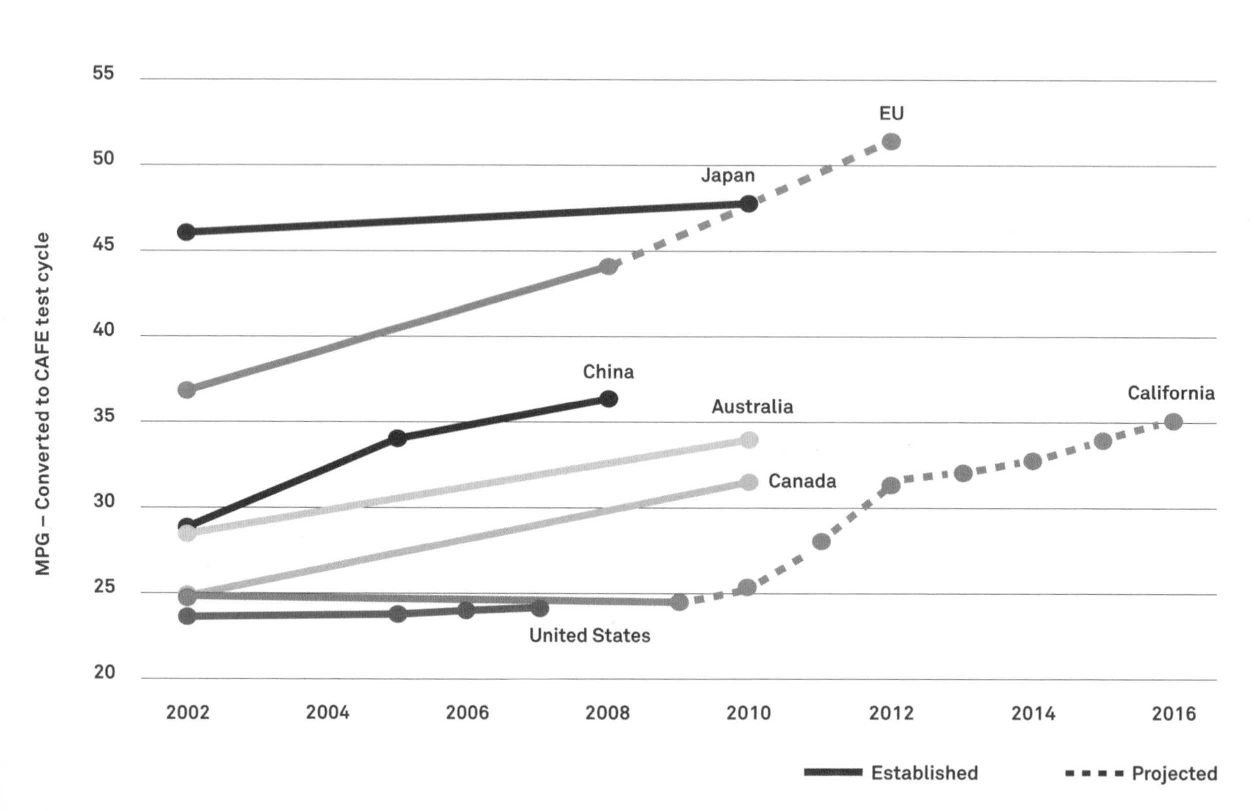

Paying attention to the environment doesn't mean businesses can't make money. There are successful automobile companies building fuel-efficient cars. Unfortunately, they are not American companies. Japanese companies like Toyota and Honda are growing while U.S. auto companies like Ford and General Motors are in deep trouble. The graph shows just how poorly these U.S. companies fared (the orange bars) in the months from February to November in 2005. They keep trying to sell large, inefficient gas-guzzlers even though fewer and fewer people are buying them.

And after California took steps to set higher mileage standards for cars sold there, what did U.S. car companies do? They sued.

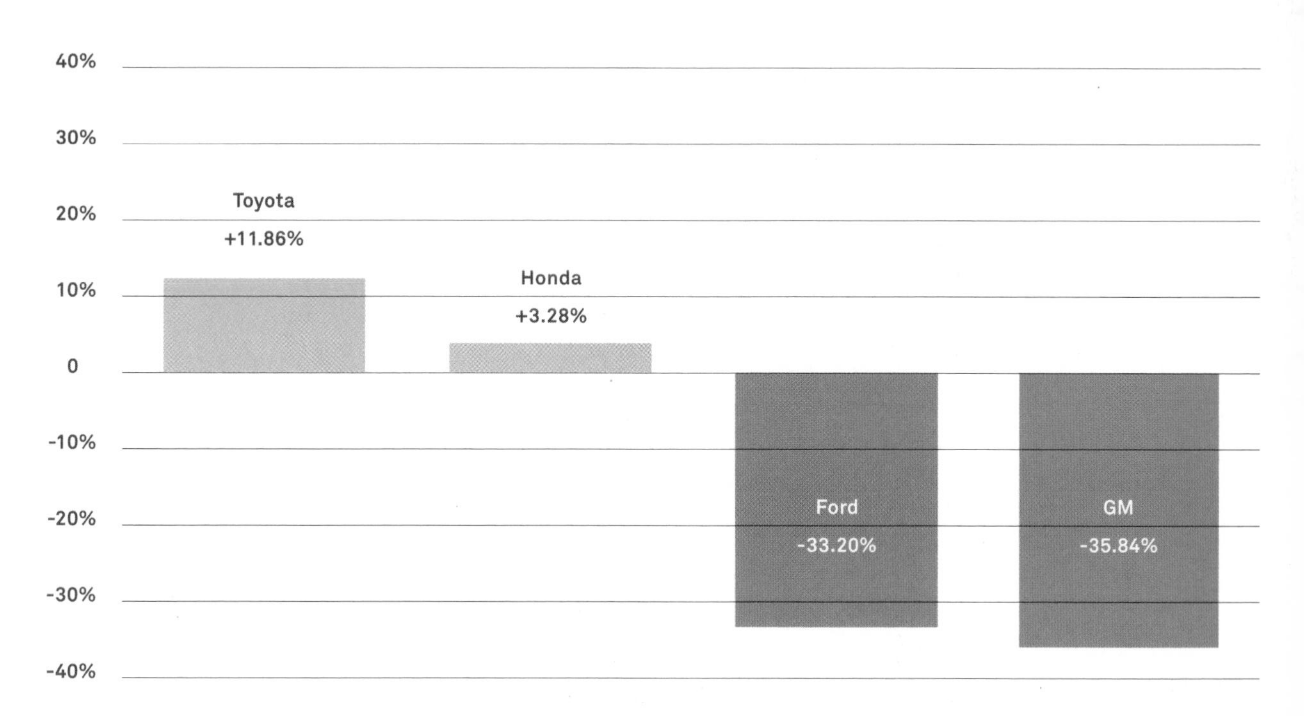

chapter fifteen

危机
Crisis = Opportunity

ANOTHER BIG PROBLEM with global warming is that an astonishing number of people go straight from denial to despair, without pausing at the step in between. Yes, there's a crisis . . . but we can do something about it.

Luckily, more and more businesses are going in the right direction. Individual families are, too. We are constantly developing new technology that can help fight global warming.

Solar Panels

On a bright sunny day, the sun shines approximately 1,000 watts of energy per square yard. Solar panels can collect that energy and turn it into electricity to power homes and offices.

Geothermal Power Stations

The heat stored in the earth can produce electricity. Geothermal power stations are now generating electricity this way. They can be built wherever there is high-temperature groundwater near the surface.

SOLAR PANELS

GEOTHERMAL POWER STATION

COMPACT FLUORESCENT LIGHTBULBS

GREEN ROOF

Fluorescent Lightbulbs

A "normal" lightbulb (also known as an incandescent lightbulb) produces a lot of heat in addition to light. The heat is a waste of energy. A fluorescent bulb uses a different, much more efficient method to produce light. You can buy a 15-watt fluorescent bulb that will produce the same amount of light as a 60-watt incandescent bulb.

Green Roofs

A green roof is one that is built to support plant life. Green roofs not only filter pollutants such as CO_2 from the air, they can also provide food, reduce heat in densely populated areas of cities, insulate buildings—and they are beautiful.

Hybrid Cars

Hybrid cars don't depend on gas alone to move. They also use batteries. Most commonly, hybrids use electric batteries to power the engine. The Toyota Prius is one example of a hybrid car.

Hydrogen Fuel-Cell Buses

Depending only partially on gas, fuel-cell hybrid buses offer mass transportation that is better for the environment. For example, buses that use hydrogen cells emit exhaust that is water vapor instead of CO_2. Certain cities in California have had successful trial runs of these vehicles.

HYBRID CAR

HYDROGEN FUEL CELL

Wind Power

Without wind power, some pioneers might never have settled in the Great Plains. It was the windmill that for generations tirelessly pumped underground water to the surface, helping settlers cook, wash, and tend their livestock.

Today, utility companies are investing in gigantic wind farms. A 100-megawatt wind farm (fifty 300-foot towers carrying mammoth turbines) can power 24,000 homes. You would have to burn 50,000 tons of coal—releasing vast amounts of CO_2—to provide the same amount of electricity.

Middelgrunden offshore wind farm, Copenhagen, Denmark, 2001

So many other countries are doing far more than we are. All of the nations listed below have ratified the Kyoto Protocol, a treaty that sets targets for limiting greenhouse gases. Only two advanced nations haven't ratified it. Australia is one . . . and the United States is the other.

RATIFIED BY

Albania
Algeria
Antigua
Arabia
Argentina
Armenia
Austria
Azerbaijan
Bahamas
Bahrain
Bangladesh
Barbados
Belarus
Barbuda
Belgium
Belize
Benin
Bhutan
Bolivia
Botswana
Brazil
Bulgaria
Burkina Faso
Burundi
Cambodia
Cameroon
Canada
Cape Verde
Chile
China
Colombia
Cook Islands
Costa Rica
Cuba
Cyprus
Czech Republic
Democratic Republic of Congo
Denmark
Djibouti
Dominica
Dominican Republic
Ecuador
Egypt
El Salvador
Equatorial Guinea
Eritrea
Estonia
Ethiopia
Fiji
Finland
France
Gambia
Georgia
Germany
Ghana
Greece
Grenada
Guatemala
Guinea
Guinea-Bissau
Guyana
Haiti
Honduras
Hungary
Iceland
India
Indonesia
Iran
Ireland
Israel
Italy
Jamaica
Japan
Jordan
Kenya
Kiribati
Kuwait
Kyrgyzstan
Laos
Latvia
Lesotho
Liberia
Liechtenstein
Lithuania
Luxembourg
Macedonia

Madagascar
Malawi
Malaysia
Maldives
Mali
Malta
Marshall Islands
Mauritania
Mauritius
Mexico
Micronesia
Moldova
Monaco
Mongolia
Morocco
Mozambique
Myanmar
Namibia
Nauru
Nepal
Netherlands
New Zealand
Nicaragua
Niger
Nigeria
Niue
North Korea
Norway
Oman
Pakistan
Palau
Panama
Papua New Guinea
Paraguay
Peru
Philippines
Poland
Portugal
Qatar
Romania
Russian Federation
Rwanda
Saint Lucia
Saint Vincent & Grenadines
Samoa
Saudi Arabia
Senegal
Seychelles
Singapore
Slovakia
Slovenia
Solomon Islands
South Africa
South Korea
Spain
Sri Lanka
Sudan
Swaziland
Sweden
Switzerland
Syrian Arab Republic
Tanzania
Thailand
Togo
Trinidad & Tobago
Tunisia
Turkmenistan
Tuvalu
United Arab Emirates
Uganda
Ukraine
United Kingdom
Uruguay
Uzbekistan
Vanuatu
Venezuela
Vietnam
Yemen
Zambia

NOT RATIFIED BY

Australia
United States

Many U.S. cities are abiding by the Kyoto treaty on their own and even going beyond the Kyoto measures to reduce global warming pollution. Here is a list of them.

ARKANSAS
Fayetteville
Little Rock
North Little Rock

CALIFORNIA
Albany
Aliso Viejo
Arcata
Berkeley
Burbank
Capitola
Chino
Cloverdale
Cotati
Del Mar
Dublin
Fremont
Hayward
Healdsburg
Hemet
Irvine
Lakewood
Los Angeles
Long Beach
Monterey Park
Morgan Hill
Novato
Oakland
Palo Alto
Petaluma
Pleasanton
Richmond
Rohnert Park
Sacramento
San Bruno
San Francisco
San Luis Obispo
San Jose
San Leandro
San Mateo
Santa Barbara
Santa Cruz
Santa Monica
Santa Rosa
Sebastopol
Sonoma
Stockton
Sunnyvale
Thousand Oaks
Vallejo
West Hollywood
Windsor

COLORADO
Aspen
Boulder
Denver
Telluride

CONNECTICUT
Bridgeport
Easton
Fairfield
Hamden
Hartford
Mansfield
Middletown
New Haven
Stamford

DELAWARE
Wilmington

FLORIDA
Gainesville
Hallandale Beach
Holly Hill
Hollywood
Key Biscayne
Key West
Lauderhill
Miami
Miramar
Pembroke Pines
Pompano Beach
Port St. Lucie
Sunrise
Tallahassee
Tamarac
West Palm Beach

GEORGIA
Atlanta
Athens
East Point
Macon

HAWAII
Hilo
Honolulu
Kauai
Maui

ILLINOIS
Carol Stream
Chicago
Highland Park
Schaumburg
Waukegan

INDIANA
Columbus
Fort Wayne
Gary
Michigan City

IOWA
Des Moines

KANSAS
Lawrence
Topeka

KENTUCKY
Lexington
Louisville

LOUISIANA
Alexandria
New Orleans

MARYLAND
Annapolis
Baltimore
Chevy Chase

MASSACHUSETTS
Boston
Cambridge
Malden
Medford
Newton
Somerville
Worcester

MICHIGAN
Ann Arbor
Grand Rapids
Southfield

MINNESOTA
Apple Valley
Duluth
Eden Prairie
Minneapolis
St. Paul

MISSOURI
Clayton
Florissant
Kansas City
Maplewood
St. Louis
Sunset Hills
University City

MONTANA
Billings
Missoula

NEBRASKA
Bellevue
Lincoln
Omaha

NEVADA
Las Vegas

NEW HAMPSHIRE
Keene
Manchester
Nashua

NEW JERSEY
Bayonne
Bloomfield
Brick Township
Elizabeth
Hamilton
Hightstown
Hope
Hopewell
Kearny
Newark
Plainfield
Robbinsville
Westfield

NEW MEXICO
Albuquerque

NEW YORK
Albany
Buffalo
Hempstead
Ithaca
Mt. Vernon
New York City
Niagara Falls
Rochester
Rockville Centre
Schenectady
White Plains

NORTH CAROLINA
Asheville
Chapel Hill
Durham

OHIO
Brooklyn
Dayton
Garfield Heights
Middletown
Toledo

OKLAHOMA
Norman North

OREGON
Corvallis
Eugene
Lake Oswego
Portland

PENNSYLVANIA
Erie
Philadephia

RHODE ISLAND
Pawtucket
Providence
Warwick

SOUTH CAROLINA
Charleston
Sumter

TEXAS
Arlington
Austin
Denton
Euless
Hurst
Laredo
McKinney

UTAH
Moab
Park City
Salt Lake City

VERMONT
Burlington

VIRGINIA
Alexandria
Charlottesville
Virginia Beach

WASHINGTON
Auburn
Bainbridge Island
Bellingham
Burien
Edmonds
Issaquah
Kirkland
Lacey
Lynnwood
Olympia
Redmond
Renton
Seattle
Tacoma
Vancouver

WASHINGTON, DC

WISCONSIN
Ashland
Greenfield
La Crosse
Madison
Racine
Washburn
Wauwatosa
West Allis

We've Done It Before

In the 1980s, the problem of the hole in the ozone layer was said to be impossible to fix. After all, the causes were global and the solution required cooperation from every nation in the world. But the United States took the lead. We drafted a treaty, secured worldwide agreement on it, and began to eliminate the chemicals (chlorofluorocarbons, or CFCs) that were causing the problem. We took action then; we must do so now.

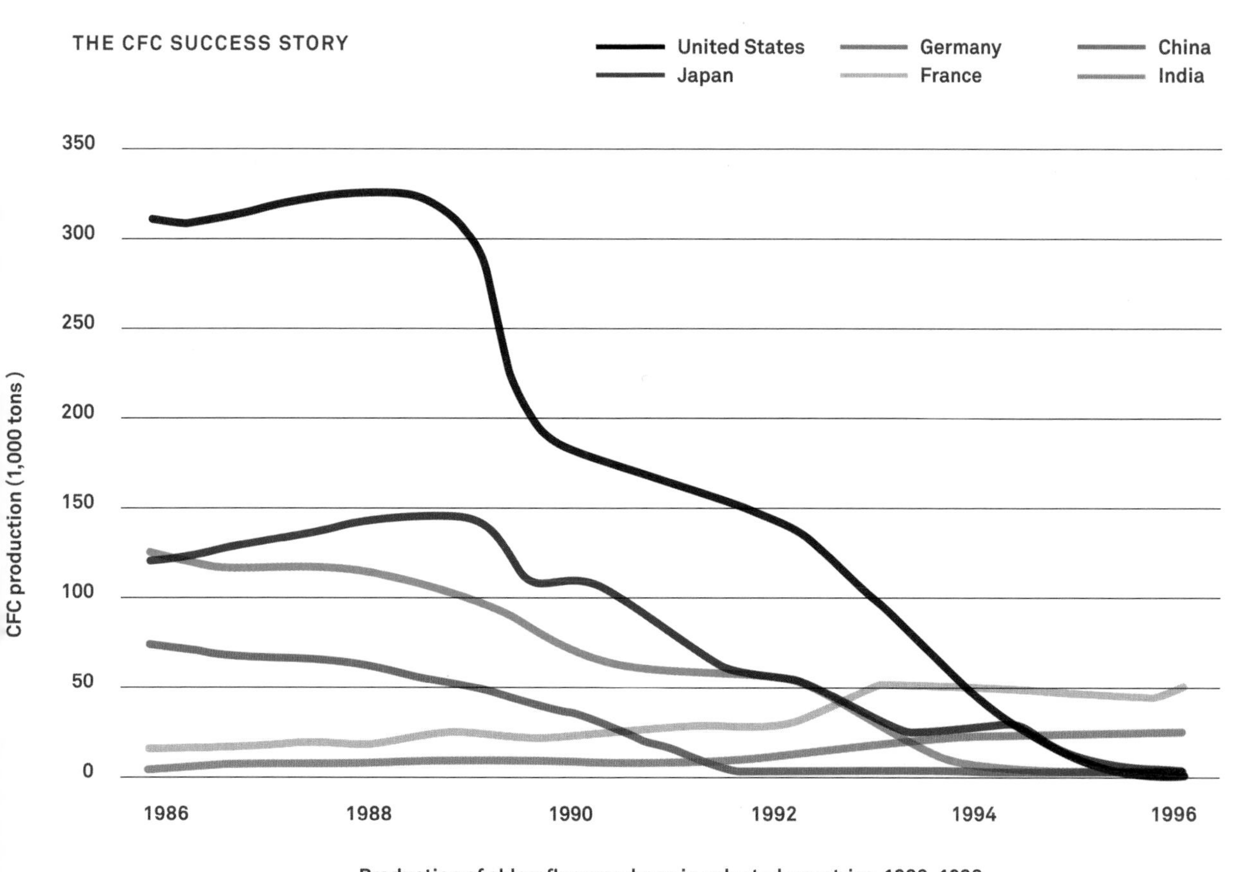

The problem of global warming is harder to solve. It demands more of us.

The climate crisis presents us with an inconvenient truth. It means we are going to have to change the way we live our lives. Whether these changes involve something as minor as using different lightbulbs, or as major as switching from oil and coal to other fuels, they will require effort and cost money. But many of these needed changes will actually save money and make us more efficient and productive. We all must take action so that our democracy creates laws to protect our planet, because we simply can't afford not to act.

View of star-forming region S106 IRS4 as seen from the Subaru Telescope, Mauna Kea, Hawaii, 2001

The twenty-first century is your century—and you can make it a time of renewal by seizing the opportunities that are bound up in this crisis. You can preserve the health of our planet so that it will remain beautiful for generations and centuries to come.

This picture of the universe was taken from a spacecraft that had traveled four million miles beyond our solar system. The pale blue dot, visible in the center of the band of light is us. Earth. Carl Sagan suggested to NASA that the photo be taken. Later he spoke about how all the triumphs and tragedies—all the wars and famines as well as all the miracles of sciences and the beauty of human creativity—have happened on this pale blue dot.

It is our only home.
And we must take care of it.

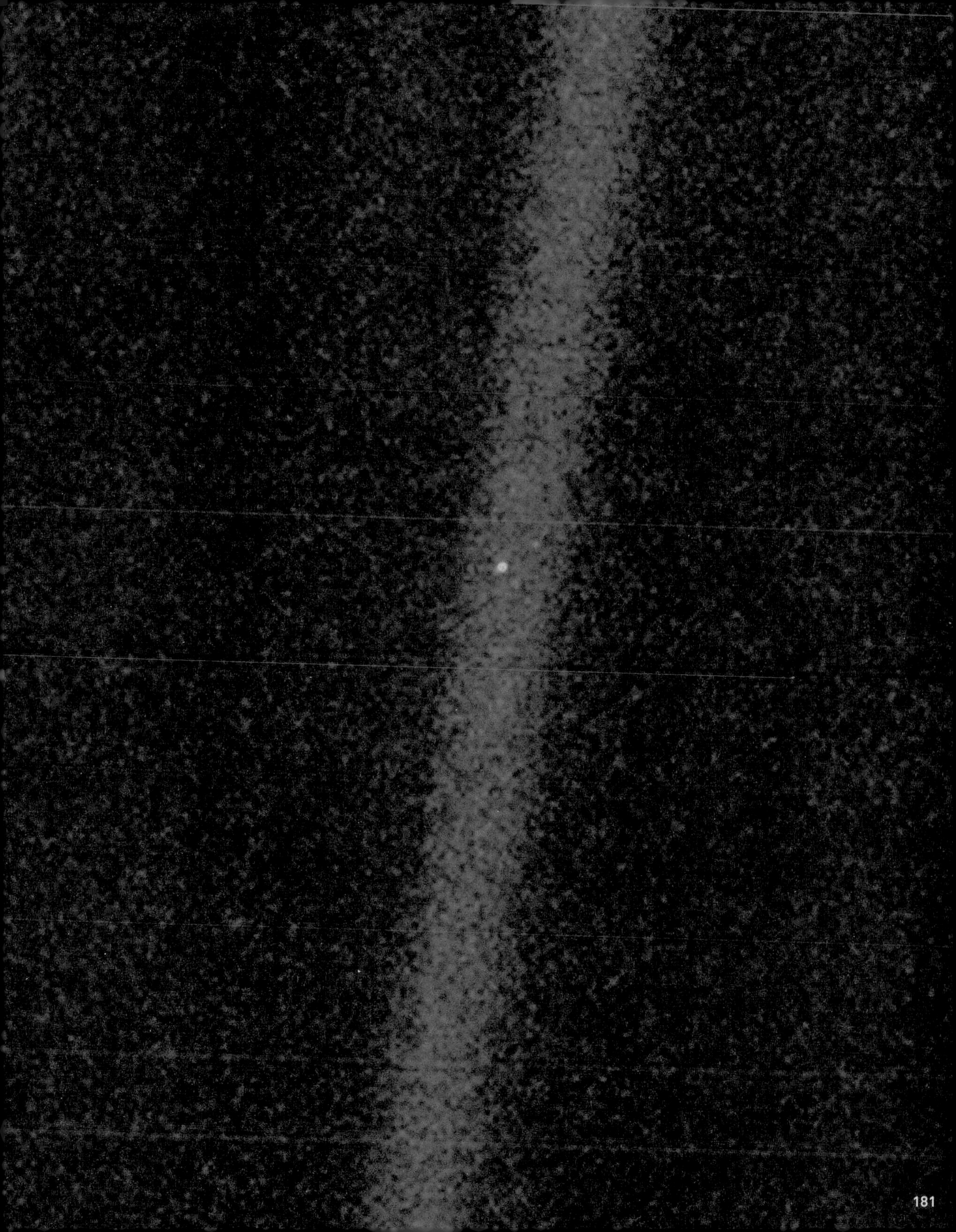

Take Action

We have everything we need to begin solving the climate crisis.

Each one of us is a cause of global warming, but each of us can become part of the solution: in the decisions we make on what we buy, the amount of electricity we use, the car our family drives, and how we live our lives.

We can make choices so that no unnecessary carbon is released because of our own individual actions.

Here are a few examples of actions you can take to make a difference:

—Avoid overpackaged food and other products with extra wrappers and layers of unnecessary plastic. For instance, use a refillable water bottle instead of buying single-use plastic bottles.

—Ride your bike or walk instead of riding in a car.

—When you're at home, remember to turn off any lights that you aren't using. And don't stand in front of an open refrigerator door—leaving it open for just a few seconds wastes a lot of energy.

—Talk to your parents, teachers, and friends about the impact they might be having on the environment. Explain what they can do to help solve the climate crisis.

For additional information, you can go to www.climatecrisis.net.

ACKNOWLEDGMENTS

My wife, Tipper, began urging me to write this book several years ago, arguing that public concern and curiosity about global warming had advanced considerably since I published *Earth in the Balance* in early 1992, and that the public interest would be well served by a new kind of book that combined fresh and up-to-date textual analysis with pictures and graphic images that would make the climate crisis more accessible and less forbidding to a wider audience. And as has so often been the case in our 36 years of marriage, she was not only right, but was patiently and persistently right for a considerable period of time before I *realized* she was right. In any case, she then helped me at every stage of the process to make the idea for the book a reality. Without Tipper, needless to say, this book never would have existed.

After I finally completed the text at the end of 2005, the two of us assembled it along with all the pictures and graphics in the proper order and shipped it from our home in Nashville to my agent, Andrew Wylie, in New York City, on New Year's Eve, just before midnight. And Andrew, as usual, knew exactly how to put the manuscript in just the right hands—in a way that made sure that it was given the best chance to be made into the book that you are now holding.

My experience with Rodale has been nothing short of spectacular. Steve Murphy, the CEO, made this project a personal cause, and moved heaven and earth to complete a complex and unusual project in a beautiful way, in record time. I'd also like to thank the Rodale family, whose lifelong commitment to the environment is inspirational, and whose generous support for this project is greatly appreciated.

I am especially grateful to my editor, Leigh Haber, for her indispensable role in shaping this book, editing it with such skill, for her suggestions and creative ideas—and for making the whole process fun from start to finish, even as we were all working at breakneck speed to meet the impossibly tight deadlines. Thanks also to everyone else at Rodale who worked so hard on this project: Liz Perl and her team, Tami Booth Corwin, Caroline Dube, Mike Sudik and his great production team, Andy Carpenter and his dedicated team, and Chris Krogermeier and her staff.

Thanks also to the team at Penguin, including Doug Whiteman, Regina Hayes, Catherine Frank, Janet Pascal, Jim Hoover, and Kendra Levin.

Special thanks to Jane O'Connor.

I am also grateful to Leigh for her decision to invite Charlie Melcher and his wonderful and dedicated colleagues at Melcher Media and mgmt. design to become part of the extraordinary creative team that Rodale organized and that Leigh headed. A special thank you for many late nights to Jessi Rymill, Alicia Cheng, and Lisa Maione. Thanks also to Bronwyn Barnes, Duncan Bock, Jessica Brackman, David Brown, Nick Carbonaro, Stephanie Church, Bonnie Eldon, Rachel Griffin, Eleanor Kung, Kyle Martin, Patrick Moos, Erik Ness, Abigail Pogrebin, Lia Ronnen, Hillary Rosner, Alex Tart, Shoshana Thaler, and Matt Wolf. Charlie and his group have brought an extremely creative approach and a truly impressive work ethic to designing and producing this complex presentation.

In addition, I would like to thank Mike Feldman and his colleagues at the Glover Park Group for their help. The book and the movie have been separate projects, but the movie team deserves special thanks for all of the many things they have done to facilitate the success of this book, even as the movie was in its final stages of preparation. Thanks especially to:

Lawrence Bender
Scott Z. Burns
Lesley Chilcott
Megan Colligan
Laurie David
Davis Guggenheim
Jonathan Lesher
Jeff Skoll

Very special thanks to my dear friend Natilee Dunning for her insightful suggestions and skillful editing assistance during the completion of the essays.

Special thanks to Matt Groening.

My friend Melissa Etheridge was incredibly responsive and helpful in composing and singing an original song for the end of the movie.

And many years before there was a movie, Gary Allison and Peter Knight helped to organize an early project that turned out to be invaluable in the projects I have pursued over the past couple of years.

Thanks to Ross Gelbspan for his dedication and tirelessness.

Gail Buckland has been a terrific help in finding pictures. She is literally the most knowledgeable person in the world where photo archives are concerned, and I always enjoy working with her and learning from her.

In addition, the people at Getty Images went above and beyond the call of duty to help with this project.

Thanks are due especially to Jill Martin and Ryan Orcutt at Duarte Design—and also Ted Boda, whose place has been taken by Ryan—for all of their countless hours over the past several years helping me find images and design graphics to illustrate complicated concepts and phenomena.

The Gore family in October 2006,
BACK ROW, LEFT TO RIGHT:
Karenna Gore Schiff, Wyatt Schiff (7 years old), Drew Schiff, Anna Schiff (5 years old), Tipper Gore, and Al Gore
FRONT ROW, LEFT TO RIGHT:
Albert Gore, Sarah Gore, Kristin Gore, and Paul Cusak

Tom Van Sant has dedicated many years of his life to conceiving and painstakingly creating one of the most remarkable sets of photographic images of the Earth ever made. His images inspired me 17 years ago, when I first saw them, and he has continued to improve them ever since.

I am grateful to be able to use the latest one-meter-resolution imagery Tom has produced.

Among the many scientists who have helped me over the years to better understand these issues, I want to single out a small group that has played a particular role in advising me on this book, and the movie that is part of the overall project:

James Baker
Rosina Bierbaum
Eric Chivian
Paul Epstein
Jim Hansen
Henry Kelly
James McCarthy
Mario Molina
Michael Oppenheimer
David Sandalow
Ellen & Lonnie Thompson
Yao Tandong

In addition, three distinguished scientists whose work and inspiration were central to this book are now deceased:

Charles David Keeling
Roger Revelle
Carl Sagan

I am grateful to Steve Jobs and my friends at Apple Computer, Inc. (I am on the Board of Directors) for helping with the Keynote II software program that I have used extensively in putting this book together.

I am particularly thankful to my partners and colleagues at Generation Investment Management for help in analyzing a number of complex questions dealt with in the book. And I want to thank my colleagues at Current TV for their help in locating several images that are used in the book.

I would also like to acknowledge MDA Federal, Inc., for their help in calculating and portraying the imagery to demonstrate with scientific precision the impact of sea level rise in various cities around the world.

Throughout my work on this book, Josh Cherwin, of my staff, has been unbelievably helpful in countless ways. Also, the rest of my entire staff has contributed a tremendous amount:

Lisa Berg
Dwayne Kemp
Melinda Medlin
Roy Neel
Kalee Kreider

Several members of my family played a direct role in helping me with this project:

Karenna Gore Schiff and Drew Schiff
Kristin Gore and Paul Cusack
Sarah Gore
Albert Gore, III
and my brother-in-law, Frank Hunger

All of them have been my constant inspiration and the principal way that I personally connect to the future.

Credits

Illustrations by Michael Fornalski
Information graphics by mgmt. design

The publisher and packager wish to recognize the following individuals and organizations for contributing photographs and images to this project:

Animals Animals; ArcticNet; Yann Arthus-Bertrand (www.yannarthusbertrand.com); Buck/Renewable Films; Tracey Dixon; Getty Images; Kenneth E. Gibson; Tipper Gore; Paul Grabbhorn; Frans Lanting (www.lanting.com); Eric Lee; Mark Lynas; Dr. Jim McCarthy; Bruno Messerli; Carl Page; W. T. Pfeffer; Karen Robinson; Vladimir Romanovsky; Lonnie Thompson; and Tom Van Sant

Page 1, 2: NASA; 5: Tipper Gore; 6: courtesy of the Gore family; 11: courtesy of the Gore family; 12–13: NASA; 14–15: Tom Van Sant/GeoSphere Project; 16–17: Tom Van Sant/GeoSphere Project and Michael Fornalski; 18–19: Getty Images; 20–21: Steve Cole/Getty Images; 22–23: Derek Trask/Corbis; 24–25: Tom Van Sant/GeoSphere Project and Michael Fornalski; 28–29: Tom Van Sant/GeoSphere Project; 30–31: Tom Van Sant/GeoSphere Project and Michael Fornalski; 32: Lonnie Thompson; 33 (top): Carle Page; 33 (bottom): Bruno Messerli; 34–35: U.S. Geological Survey; 36: (all photographs) Copyright by Sammlung Gesellschaft fuer oekologische Forschung, Munich, Germany; 37: (composite) Daniel Beltra/ZUMA Press/Copyright by Greenpeace; 38–39: Daniel Garcia/AFP/Getty Images; 40–41: Map Resources; 42 and 43 (left): Lonnie Thompson; 43 (right): Vin Morgan/AFP/Getty Images; 50–51: Michaela Rehle/Reuters; 54–55: Paul S. Howell/Getty Images; 56: Philippe Colombi/Getty Images; 58–59: NOAA; 60–61: NASA; 62–63: Don Farrall/Getty Images; 64–65: Andrew Winning/Reuters/Corbis; 66–67: (composite) NASA/NOAA/Plymouth State Weather Center; 68: Marko Georgiev/Getty Images; 69 (top): David Portnoy/Getty Images; 69 (bottom): Robyn Beck/AFP/Getty Images; 70–71: Vincent Laforet/The New York Times; 73: Keystone/Sigi Tischler; 74: Sebastian D'Souza/AFP/Getty Images; 75: China Photos/Getty Images; 76–77: Tom Van Sant/GeoSphere Project and Michael Fornalski; 78 (all): NASA; 79: Stephane De Sakutin/AFP/Getty Images; 80–81: Natural Resources of Canada, 2001; 82–83: Derek Mueller and Warwick Vincent/Laval University/ArcticNet; 85: Michael Fornalski; 86–87: Tracey Dixon; 88–89: Peter Essick/Aurora/Getty Images; 90 (top): Vladimir Romanovsky/Geophysical Institute/UAF; 90 (bottom) Mark Lynas; 91 (graphic): Arctic Climate Impact Assessment; 93: Natural Resources of Canada, 2001; 94–95: Frans Lanting; 96: British Antarctic Survey; 98: map, J. Kaiser, *Science*, 2002; 98–99: all satellite images, NASA; 100–101: Frans Lanting; 103: Tom Van Sant/GeoSphere Project; 104–105: (graphic) Renewable Films/ACIA; 106: (graphic) Buck/Renewable Films and NASA; 107: Roger Braithwaite/Peter Arnold; 108–109: Mark Lynas; 110–112 (all images): MDA Federal Inc. and Brian Fisher/Renewable Films; 113: Ooms Avenhorn Groep bv; 114 (top): U.S. CIA; 114 (bottom) and 115: MDA Federal Inc. and Brian Fisher/Renewable Films; 116: Google Earth; 117: MDA Federal Inc. and Brian Fisher/Renewable Films; 118–119: Tom Van Sant/GeoSphere Project and Michael Fornalski; 120–121: Paul Nicklen/National Geographic/Getty Images; 122–123: Bill Curtsinger/National Geographic/Getty Images; 124–125: Janerik Henriksson/SCANPIX/Retna Ltd.; 127 (top): Bob Turkel/SmugMug; 127 (bottom): USDA; 131: Benelux Press/Getty Images; 133: Kenneth E. Gibson/USDA Forest Service/www.forestryimages.org; 134: (left to right, top to bottom): Juan Manuel Renjifo/Animals Animals, David Haring/OSF/Animals Animals, Rick

Price Survival/OSF/Animals Animals, Juergen and Christine Sohns/Animals Animals, Johnny Johnson/Animals Animals, Frans Lanting; Michael Fogden/OSF/Animals Animals, Johnny Johnson/Animals Animals, Raymond Mendez/Animals Animals, Leonard Rue/Animals Animals, Frans Lanting, Frans Lanting, Peter Weimann/Animals Animals, Don Enger/Animals Animals, Erwin and Peggy Bauer/Animals Animals, Frans Lanting; 136-137: Yann Arthus-Bertrand [Refuse dump in Mexico City, Mexico (19°24' N, 99°01' W). Household refuse is piling up on all continents and poses a critical problem for major urban centers, like the problem of air pollution resulting from vehicular traffic and industrial pollutants. With some 21 million residents, Mexico City produces nearly 20,000 tons of household refuse a day. As in many countries, half of this debris is sent to open dumps. The volume of refuse is increasing on our planet along with population growth and, in particular, economic growth. Thus, an American produces more than 1,500 pounds (700 kg) of domestic refuse each year, about four times more than a resident of a developing country and twice as much as a Mexican. The volume of debris per capita in industrialized nations has tripled in the past 20 years. Recycling, reuse, and reduction of packaging materials are potential solutions to the pollution problems caused by dumping and incineration, which still account for 41% and 44%, respectively, of the annual volume of household garbage in France.]; 140–141: Yann Arthus-Bertrand [Shinjuku district of Tokyo, Japan (35°42' N, 139°46' E). In 1868 Edo, originally a fishing village built in the middle of a swamp, became Tokyo, the capital of the East. The city was devastated by an earthquake in 1923 and by bombing in 1945, both times to be reborn from the ashes. Extending over 43 miles (70 km) and holding a population of 28 million, the megalopolis of Tokyo (including surrounding areas such as Yokohama, Kawasaki, and Chiba) is today the largest metropolitan region in the world. It was not built according to an inclusive urban design and thus contains several centers, from which radiate different districts. Shinjuku, the business district, is predominantly made up of an impressive group of administrative buildings, including the city hall, a 798-foot-high (243 m) structure that was modeled after the cathedral of Notre Dame in Paris. In 1800 only London had more than 1 million inhabitants; today 326 urban areas have reached that number, including 180 in developing countries and 16 megalopolises that have populations of more than 10 million. Urbanization has led to a tripling of the population living in cities since 1950.]; 142: Peter Essick/Aurora/Getty Images; 143: Kevin Schafer/Corbis; 144–145: Stephen Ferry/Liaison/Getty Images; 146–147: National Geographic; 148–149: NASA; 150–151: David Turnley/Corbis; 152–153: Beth Wald/Aurora/Getty Images USGS; 154: Palma Collection/Getty Images; 155: Corbis; 162: The New York Times; 167: (top to bottom) Mark Segal/Getty Images; James Davis/Eye Ubiquitous/Corbis; 168 (top to bottom): William Thomas Cain/Getty Images, City of Chicago; 171: David Paul Morris/Getty Images; Joe Raedle/Getty Images; 172–173: Yann Arthus-Bertrand [Middelgrunden offshore wind farm, near Copenhagen, Denmark (55°40' N, 12°38' E). Since late 2000, one of the largest offshore wind farms to date has stood in the Øresund strait, which separates Denmark from Sweden. Its 20 turbines, each equipped with a rotor 250 feet (76 m) in diameter, standing 210 feet (64 m) above the water, form an arc with a length of 2.1 miles (3.4 km). With 40 megawatts of power, the farm produces 89,000 mW annually (about 3% of the electricity consumption of Copenhagen). By 2030 Denmark plans to satisfy 40% of its electricity needs by means of wind energy (as opposed to 13% in 2001). Although renewable forms of energy still make up less than 2% of the primary energy used worldwide, the ecological advantages are attracting great interest. Thanks to technical progress, which has reduced the noise created by wind farms (installed about one-third of a mile, or 500 m, from residential areas), resistance is fading. And with a 30% average annual growth rate in the past four years, the wind farm seems to be here to stay.]; 178–179: Subaru Telescope, National Astronomical Observatory for Japan. All rights reserved; 180–181: NASA.

Index

Numbers in italic refer to illustrations

This book is printed on Appleton Green Power Utopia 80# matte. The paper is made of 30% postconsumer waste. The virgin fibers are grown under a certified forestry management system. The pulp is produced chlorine-free using 100% green power.